FUZZY LOGIC

A PRACTICAL APPROACH

LIMITED WARRANTY AND DISCLAIMER OF LIABILITY

FUZZY LOGIC

A PRACTICAL APPROACH

F. Martin McNeill
Ellen Thro

AP PROFESSIONAL

Boston   San Diego   New York
London   Sydney   Tokyo   Toronto

AP PROFESSIONAL
955 Massachusetts Avenue, Cambridge, MA 02139

An Imprint of ACADEMIC PRESS, INC.
A Division of HARCOURT BRACE & COMPANY

United Kingdom Edition published by
ACADEMIC PRESS LIMITED
24–28 Oval Road, London NW1 7DX

Library of Congress Cataloging-in-Publication Data
McNeill, F. Martin, date.
    Fuzzy logic: a practical approach / F. Martin McNeill, Ellen
Thro.
        p.   cm.
    Includes bibliographical references and index.
    ISBN 0-12-485965-8 (acid-free paper)
    1. Automatic control. 2. Expert systems (Computer science)
3. Fuzzy logic. I. Thro, Ellen. II. Title.
TJ213.M355   1994   94-30787
006.3'3—dc20   CIP

Printed in the United States of America
94 95 96 97 98 IP   9 8 7 6 5 4 3 2 1

Dedication of this book is to the memory of Merrill Meeks Flood, Ph.D. To the extent that the fact of existence is magic, he personified that magic.

—FMM

Acknowledgments for support go to the following:

Valerio Aisa, Merloni Eletrodomestica spa, Viale Aristide Merloni 45, 60044 Fabriano (AN) Italy.

and

David Brubaker—the Huntington Group
David Crumpton—the Motorola Semiconductors, Inc.
Dr. Michael O'Hagan—Fuzzy Logic, Inc.
Derek Stubbs—Advanced Forecasting Technologies

CONTENTS

Psychology
artificial intelligence

AI

~~Biological~~

neural nets
Genetic algorithm
evolutionary progr

Intelligent Technology
- Uncertainty
   - Probability
   - Fuzzy Set theory → Complex system models
         transform natural language to mathematical formula

- Natural language

FOREWORD

The last decade has seen a large interest in technologies that have as their motivation some aspect of human function. Some of these, like artificial intelligence, can be seen to be rooted in the psychological domain. Others, like neural networks, genetic algorithms, and evolutionary programming, are inspired by reconsiderations of biological processes. Common to all these so-called "intelligent technologies" is a need to represent knowledge in a manner that is both faithful to the human style of processing information as well as a form amenable to computer manipulation.

Fuzzy sets were originally introduced in 1965; the related discipline of fuzzy logic is proving itself as the most appropriate medium to accomplish this task. At one level, fuzzy logic can be viewed as a language that allows one to translate sophisticated statements from natural language into a mathematical formalism. Once we have this mathematical form of knowledge, we are able to draw upon hundreds of years of recent history in technology to manipulate this knowledge.

While the original motivation was to help manage the pervasive imprecision in the world, the early practitioners of fuzzy logic dealt primarily with theoretical issues. Many early papers were devoted to basic foundations and to "potential" applications. This early phase was also marked by a strong need to distinguish fuzzy logic from probability theory. As is well understood now, fuzzy set theory and probability theory are directed at different types of uncertainty. The next phase of the development of the discipline was driven by the success, particularly in Japan, of using fuzzy logic to design simple controllers. This success has sparked a worldwide interest in using this technology for the construction of complex systems models in engineering disciplines.

With the publication of this book we are beginning to see the emergence of the next phase of fuzzy logic. During this phase we will see the opening of the power of this methodology to middle-level "technocrats." In addition, the focus of this book, rather than being strictly on engineering problems, provides a number of broader applications. The authors are to be complimented on providing a book that will be very useful to those who desire to *use* fuzzy logic to solve their problems. The book has many examples and complementary software to help the novice.

I look forward to a future in which the techniques of fuzzy logic will become as pervasive on desktop computers as spreadsheets and databases. The authors of this book have taken an important step in helping realize this future.

Ronald R. Yager
New York
June 1994

CHAPTER 1

THE FUZZY WORLD

What's the process of parallel parking a car?

First you line up your car next to the one in front of your space. Then you angle the car back into the space, turning the steering wheel slightly to adjust your angle as you get closer to the curb. Now turn the wheel to back up straight and—nothing. Your rear tire's wedged against the curb.

OK. Go forward slowly, steering toward the curb until the rear tire straightens out. Fine—except, you're too far from the curb. Drive back and forth again, using shallower angles.

Now straight forward. Good, but a little too close to the car ahead. Back up a few inches. Thunk! Oops, that's the bumper of the car in back. Forward just a few inches. Stop! Perfect!! Congratulations. You've just parallel-parked your car.

And you've just performed a series of fuzzy operations.

Not fuzzy in the sense of being confused. But fuzzy in the real-world sense, like "going forward slowly" or "a bit hungry" or "partly cloudy"—the distinctions that people use in decision-making all the time, but that computers and other advanced technology haven't been able to handle.

What kind of problems? For one, waiting for an elevator at lunch hour. How do you program elevators so that they pick up the most people in the least amount of time? Or how do you program elevators to minimize the waiting time for the most people?

Suppose you're operating an automated subway system. How do you program a train to start up and slow down at stations so smoothly that the passengers hardly notice?

For that matter, how can you program a brake system on an automobile so that it works efficiently, taking road and tire conditions into account?

Perhaps you have a manufacturing process that requires a very steady temperature over a many hours. What's the most efficient and reliable method for achieving it?

Or, suppose you're filming an unpredictable and fast-moving event with your camcorder—say, a birthday party of 10 three-year-olds. What kind of a camera lets you move with the action and still end up with a very nonjerky image when you play it back?

Or, take a problem far from the realm of manufacturing and engineering, such as, how do you define the term *family* for the purposes of inclusion in health insurance policy?

Do all these situations have something in common? For one thing, they're all complex and dynamic. Also, like parallel parking, they're more easily characterized by words and shades of meaning than by mathematics.

In this book you'll be immersed in the fuzzy world, not an easy process. You'll meet the basics, manipulate the tools (simple and complex), and use them to solve real-world problems. You can make your experience interactive and hands on with a series of programs on the accompanying disk. (See the Preface for an explanation of how to load it onto your hard disk.) To make the trip easier, you'll be following in the many footsteps of our fuzzy field guide, Dr. Fuzzy. The good doctor will be on call through Help menus and will show up in the book chapters with hints, further information, and encouraging messages.

E-MAIL FROM DR. FUZZY   The real world is up and down, constantly moving and changing, and full of surprises. In other words, fuzzy.
    Fuzzy techniques let you successfully handle real-world situations.

APPLES, ORANGES, OR IN BETWEEN?

As the fiber-conscious Dr. Fuzzy has discovered, one of the easiest ways to step into the fuzzy world is with a simple device found in most homes—a bowl of fruit. Conventional computers and simple digital control systems follow the either-or system. The digit's either zero or one. The answer's either yes or no. And the fruit bowl (or database cell) contains either apples or oranges.

Take Figure 1.1, for example. Is this a bowl of oranges? The answer is No.

How about Figure 1.2? Is it a bowl of oranges? The answer in this case is Yes.

This is an example of crisp logic, adequate for a situation in which the bowl does contain *either* totally apples *or* totally oranges. But life is often more complex. Take the case of the bowl in Figure 1.3. Someone has made a switch,

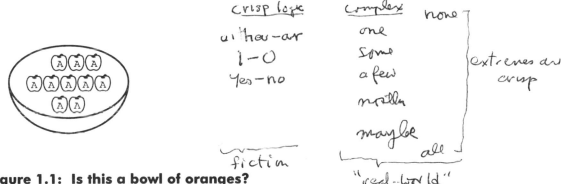

Figure 1.1: **Is this a bowl of oranges?**

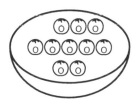

Figure 1.2: **Is *this* a bowl of oranges?**

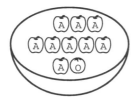

Figure 1.3: "Thinking fuzzy" about a bowl of oranges.

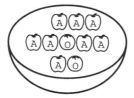

Figure 1.4: Fuzzy bowl of apples.

Figure 1.5: Fuzzy bowl of apples (continued).

swapping an orange for one of the apples in the Yes—Apple bowl. Is it a bowl of oranges?

Suppose another apple disappears, only to be replaced by an orange (Figure 1.4). The same thing happens again (Figure 1.5). And again (Figure 1.6). Is the bowl now a bowl of oranges? Suppose the process continues

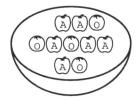

Figure 1.6: Fuzzy bowl of apples (continued).

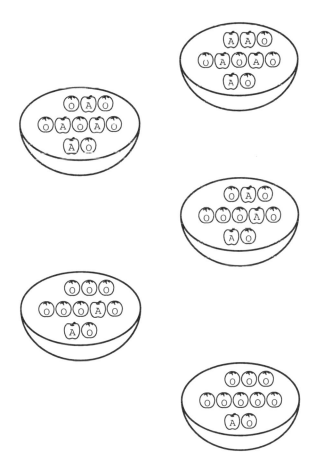

Figure 1.7: Fuzzy bowl of apples (continued).

(Figure 1.7). At some point, can you say that the "next bowl" contains oranges rather than apples?

This isn't a situation where you're unable to say Yes or No because you need more information. You have all the information you need. The situation itself makes *either* Yes *or* No inappropriate. In fact, if you had to say Yes or No, your answer would be less precise that if you answered One, or Some, or A Few, or Mostly—all of which are fuzzy answers, somewhere in between Yes and No. They handle the actual ambiguity in descriptions or presentations of reality.

Other ambiguities are possible. For example, if the apples were coated with orange candy, in which case the answer might be Maybe. The complexity of reality leads to truth being stranger than fiction. Fuzzy logic holds that crisp (0/1) logic is often a fiction. Fuzzy logic actually contains crisp logic as an extreme.

E-MAIL FROM DR. FUZZY

Really want to think fuzzy apples and oranges? They have less distinct boundaries than you might think.

Both apples and oranges are spheres, and both are about the same size. Both grow on trees that reproduce similarly. You can make a tasty drink from each. They even go to their rewards the same way, by being eaten and digested by people, or by being composted by my relatives, near and distant. If the apples are red, even the colors are related—

red + yellow = orange

And don't neglect the bowl. Both fruits nestle the same way in the same kind of bowl, and they leave similar amounts of unoccupied space.

With fuzzy logic the answer is Maybe, and its value ranges anywhere from 0 (No) to 1 (Yes).

**E-MAIL
FROM
DR. FUZZY**
Crisp sets handle only 0s and 1s.
Fuzzy sets handle all values between 0 and 1.

Crisp
No   Yes

Fuzzy

No   Slightly   Somewhat   Sort Of   A Few   Mostly   Yes, Absolutely

Looking at the fruit bowls again (Figure 1.8), you might assign these fuzzy values to answer the question, Is this a bowl of oranges?

**E-MAIL
FROM
DR. FUZZY**
Characteristics of fuzziness:
- Word based, not number based. For instance, *hot,* not *85°.*
- Nonlinear changeable.
- Analog (ambiguous), not digital (Yes/No).

If you really look at the way we make decisions, even the way we use computers and other machines, it's surprising that fuzziness isn't considered the *ordinary* way of functioning. Why isn't it? It all started with Aristotle (and his buddies).

IS THERE LIFE BEYOND MATH?

The either-apples–or-oranges system is known as "crisp" logic. It's the logic developed by the fourth century B.C. Greek philosopher Aristotle and is often called *Aristotelian* in his honor. Aristotle got his idea from the work of an earlier Greek philosopher, Pythagoras, and his followers, who believed that matter was essentially numerical and that the universe could be defined as numerical relationships. Their work is traditionally credited with providing

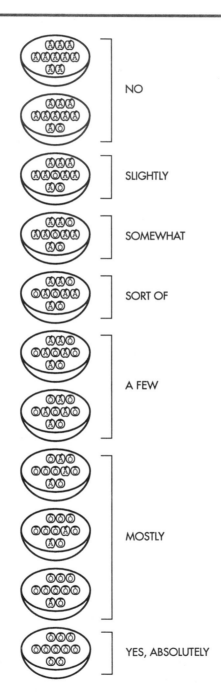

Figure 1.8: Fuzzy values.

the foundation of geometry and Western music (through the mathematics of tone relationships).

Aristotle extended the Pythagorean belief to the way people think and make decisions by allying the precision of math with the search for truth. By the tenth century A.D., Aristotelian logic was the basis of European and Middle Eastern thought. It has persisted for two reasons—it simplifies thinking about problems and makes "certainty" (or "truth") easier to prove and accept.

Vague Is Better

In 1994 fuzziness is the state of the art, but the idea isn't new by any means. It's gone under the name fuzzy for 25 years, but its roots go back 2,500 years. Even Aristotle considered that there were degrees of true–false, particularly in making statements about possible future events. Aristotle's teacher, Plato, had considered degrees of membership. In fact, the word *Platonic* embodies his concept of an intellectual ideal—for instance, of a chair—that could be realized only partially in human or physical terms. But Plato rejected the notion.

Skip to eighteenth century Europe, when three of the leading philosophers played around with the idea. The Irish philosopher and clergyman George Berkeley and the Scot David Hume thought that each concept has a concrete core, to which concepts that resemble it in some way are attracted. Hume in particular believed in the logic of common sense—reasoning based on the knowledge that ordinary people acquire by living in the world.

In Germany, Immanuel Kant considered that only mathematics could provide clean definitions, and many contradictory principles could not be resolved. For instance, matter could be divided infinitely, but at the same time could not be infinitely divided.

That particularly American school of philosophy called *pragmatism* was founded in the early years of this century by Charles Sanders Peirce, who stated that an idea's meaning is found in its consequences. Peirce was the first to consider "vagueness," rather than true–false, as a hallmark of how the world and people function.

The idea that "crisp" logic produced unmanageable contradictions was picked up and popularized at the beginning of the twentieth century by the flamboyant English philosopher and mathematician, Bertrand Russell.

He also studied the vagueness of language, as well as its precision, concluding that vagueness is a matter of degree.

E-MAIL FROM DR. FUZZY

Crisp logic has always had fuzzy edges in the form of paradoxes. One example is the apples–oranges question earlier in the chapter. Here are some ancient Greek versions:

- How many individual grains of sand can you remove from a sandpile before it isn't a pile any more (Zeno's paradox)?
- How many individual hairs can fall from a man's head before he becomes bald (Bertrand Russell's paradox)?

In ancient, politically incorrect mainland Greece they said, "All Cretans are liars. When a Cretan says that he's lying, is he telling the truth?" The logical problem: How stable is the idea of truth and falsity?

In the early twentieth century, Bertrand Russell (who seemed to be amazingly interested in human fuzz) asked: A man who's a barber advertises "I shave all men and only those who don't shave themselves." Who shaves the barber?

The down-home illustration involved this logical question: Can a set contain itself?

The German philosopher Ludwig Wittgenstein studied the ways in which a word can be used for several things that really have little in common, such as a *game*, which can be competitive or noncompetitive.

The original (0 or 1) set theory was invented by the nineteenth century German mathematician Georg Kantor. But this "crisp" set has the same shortcomings as the logic it's based on. The first logic of vagueness was developed in 1920 by the Polish philosopher Jan Lukasiewicz. He devised sets with possible membership values of 0, ½, and 1, later extending it by allowing an infinite number of values between 0 and 1.

Later in the twentieth century, the nature of mathematics, real-life events, and complexity all played roles in the examination of crispness. So did the amazing discovery of physicists such as Albert Einstein (relativity) and Werner Heisenberg (uncertainty). Einstein was quoted as saying, "As far as the laws of mathematics refer to reality, they are not certain, and as far as they are certain, they do not refer to reality."

The next big step forward came in 1937, at Cornell University, where Max Black considered the extent to which objects were members of a set, such as a chairlike object in the set Chair. He measured membership in degrees of usage and advocated a general theory of "vagueness."

The work of these nineteenth and twentieth century thinkers provided the grist for the mental mill of the founder of fuzzy logic, an American named Lotfi Zadeh.

Discovering Fuzziness

In the 1960s, Lotfi Zadeh invented fuzzy logic, which combines the concepts of crisp logic and the Lukasiewicz sets by defining graded membership. One of Zadeh's main insights was that mathematics can be used to link language and human intelligence. Many concepts are better defined by words than by mathematics, and fuzzy logic and its expression in fuzzy sets provide a discipline that can construct better models of reality.

E-MAIL FROM DR. FUZZY   Lotfi Zadeh says that fuzziness involves possibilities. For instance, it's possible that 6 is a large number, while it's impossible that 1 or 2 are large numbers. In this case, a fuzzy set of possible large numbers includes 3, 4, 5, and 6.

Daniel Schwartz, an American fuzzy logic researcher, organized fuzzy words under several headings. *Quantification* terms include all, most, many, about half, few, and no. *Usuality* includes always, frequently, often, occasionally, seldom, and never. *Likelihood* terms are certain, likely, uncertain, unlikely, and certainly not.

How do you "think fuzzy" about a fuzzy word—also called a *linguistic variable*—in contrast to "thinking crisp"? Dimiter Driankov and several colleagues in Germany have pointed out three ways that highlight the difference.

Suppose the variable is *largeness*. Someone gives you the number 6 and says, "6 is a large number. Do you agree or disagree?"

Figure 1.9: A threshold person either agrees or disagrees.

If you're a *threshold person*, you will flatly state either "I agree" or "I disagree." This can be drawn as in Figure 1.9.

An *estimator* will take a different approach, saying "I agree partially" (Figure 1.10). The answer may depend on the context in which the question is asked. The person might partly agree that 6 is a large number if the next number is 0.05. But if the next one is 50, then the person might disagree partially or totally.

A *conservative* takes still another approach, possibly saying, "I agree," "I disagree," or "I'm not sure." Public opinion polls often use this method. For instance, if the statement is "Are you willing to pay higher taxes to build more playgrounds"? Someone might answer, "I am if the playgrounds will help reduce juvenile crime."

Are any of these answers fuzzy? The threshold person has given a crisp answer—all or nothing. The other two people have given fuzzy ones. The estimator's answer involves a degree, so that there can be as many different responses as there are people answering the question. The conservative's answer recognizes that some questions by their nature may always have uncertain aspects or involve balancing tradeoffs.

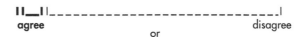

Figure 1.10: An estimator may agree partially.

THE USES OF FUZZY LOGIC

Fuzzy systems can be used for estimating, decision-making, and mechanical control systems such as air conditioning, automobile controls, and even "smart" houses, as well as industrial process controllers and a host of other applications.

The main practical use of fuzzy logic has been in the myriad of applications in Japan as process controllers. But the earliest fuzzy control developments took place in Europe.

FUZZY CONTROL SYSTEMS

The British engineer Ebrahim Mamdani was the first to use fuzzy sets in a practical control system, and it happened almost by accident. In the early 1970s, he was developing an automated control system for a steam engine using the expertise of a human operator. His original plan was to create a system based on Bayesian decision theory, a method of defining probabilities in uncertain situations that considers events after the fact to modify predictions about future outcomes.

The human operator adjusted the throttle and boiler heat as required to maintain the steam engine's speed and boiler pressure. Mamdani incorporated the operator's response into an intelligent algorithm (mathematical formula) that learned to control the engine. But as he soon discovered, the algorithm performed poorly compared to the human operator. A better method, he thought, might be to create an abstract description of machine behavior.

He could have continued to improve the learning controller. Instead, Mamdani and his colleagues decided to use an artificial intelligence method called a *rule-based expert system*, which combined human expertise with a series of logical rules for using the knowledge. While they were struggling to write traditional rules using the computer language Lisp, they came upon a new paper by Lotfi Zadeh on the use of fuzzy rules and algorithms for analysis and decision-making in complex systems. Mamdani immediately decided to try fuzziness, and within a "mere week" had read Zadeh's paper and produced a fuzzy controller. As Mamdani has written, "it was 'surprising' how easy it was to design a rule-based controller" based on a combination of linguistic and mathematical variables.

In the late 1970s, two Danish engineers, Lauritz Peter Holmblad and Jens-Jurgen Østergaard, developed the first commercial fuzzy control system, for a cement kiln. They also created one for a lime kiln in Sweden, and several others.

Other Commercial Fuzzy Systems

The most spectacular fuzzy system functioning today is the subway in the Japanese city of Sendai. Since 1987, a fuzzy control system has kept the trains rolling swiftly along the route, braking and accelerating gently, gliding into stations, stopping precisely, without losing a second or jarring a passenger.

Japanese consumer product giants such as Matsushita and Nissan have also climbed aboard the fuzzy bandwagon. Matsushita's fuzzy vacuum cleaner and washing machine are found in many Japanese homes. The washing machine evaluates the load and adjusts itself to the amount of detergent needed, the water temperature, and the type of wash cycle. Tens of thousands of Matsushita's fuzzy camcorders are producing clear pictures by automatically recording the movements the lens is aimed at, not the shakiness of the hand holding it.

Sony's fuzzy TV set automatically adjusts contrast, brightness, sharpness, and color.

Nissan's fuzzy automatic transmission and fuzzy antilock brakes are in its cars.

Mitsubishi Heavy Industries designed a fuzzy control system for elevators, improving their efficiency at handling crowds all wanting to take the elevator at the same time. This system in particular captured the imagination of companies elsewhere in the world. In the United States, the Otis Elevator Company is developing its own fuzzy product for scheduling elevators for time-varying demand.

Since the Creator of Crispness, Aristotle, had a few doubts about its application to everything, it shouldn't be a surprise that other methods of dealing with instability also exist. Some of them are a couple of centuries old.

THE VALUE OF FUZZY SYSTEMS

Writing 20 years later, Ebrahim Mamdani noted that the surprise he felt about the success of the fuzzy controller was based on cultural biases in favor of conventional control theory. Most controllers use what is called the *proportional-integral-derivative* (PID) control law. This sophisticated mathematical law assumes linear or uniform behavior by the system to be controlled. Despite this simplification, PID controllers are popular because they maintain good performance by allowing only small errors, even when external disturbances occur threaten to make the system unstable.

In fact, PID controllers were held in such high repute that any alternative control method would be expected to be equally sophisticated (meaning complicated), what Mamdani calls the "cult of analyticity."

One of the "drawbacks" of fuzzy logic is that it works with just a few simple rules. In other words, it didn't fit people's expectations of what a "good" controller should be. And it certainly shouldn't be quick and easy to produce.

Despite the culture shock, fuzzy control systems caught on—faster in Japan than in the United States—because of two drawbacks of conventional controllers. First, many processes aren't linear, and they're just too complex to be modeled mathematically. Management, economic, and telecommunications systems are examples.

Second, even for the traditional industrial processes that use PID controllers, it's not easy to describe what the term *stability* means. As Mamdani has noted, the idea of requiring mathematical definition of stability has been an academic view that hasn't really been used in the workplace. There's no industry standard of "stability," and the various methods of describing it are recommendations, not requirements. In practical terms, the value of a controller is shown by prototype tests rather than stability analysis. In fact, Mamdani says, experience with fuzzy controllers has shown that they're often more robust and stable than PID controllers.

There are five types of systems where fuzziness is necessary or beneficial:

- Complex systems that are difficult or impossible to model
- Systems controlled by human experts
- Systems with complex and continuous inputs and outputs

- Systems that use human observation as inputs or as the basis for rules
- Systems that are naturally vague, such as those in the behavioral and social sciences

Advantages and Disadvantages

According to Datapro, the Japanese fuzzy logic industry is worth billions of dollars, and the total revenue worldwide is projected at about $650 million for 1993. By 1997, that figure is expected to rise to $6.1 billion. According to other sources, Japan currently is spending $500 million a year on Fuzzy Systems R&D. And it's beginning to catch on in the United States, where it all began.

Advantages of Fuzzy Logic for System Control

- Fewer values, rules, and decisions are required.
- More observed variables can be evaluated.
- Linguistic, not numerical, variables are used, making it similar to the way humans think.
- It relates output to input, without having to understand all the variables, permitting the design of a system that may be more accurate and stable than one with a conventional control system.
- Simplicity allows the solution of perviously unsolved problems.
- Rapid prototyping is possible because a system designer doesn't have to know everything about the system before starting work.
- They're cheaper to make than conventional systems because they're easier to design.
- They have increased robustness.
- They simplify knowledge acquisition and representation.
- A few rules encompass great complexity.

Its Drawbacks

- It's hard to develop a model from a fuzzy system.
- Though they're easier to design and faster to prototype than conventional control systems, fuzzy systems require more simulation and fine tuning before they're operational.
- Perhaps the biggest drawback is the cultural bias in the United States in favor of mathematically precise or crisp systems and linear models for control systems.

FUZZY DECISION-MAKING

Fuzzy decision-making is a specialized, language oriented fuzzy system used to make personal and business management decisions, such as purchasing cars and appliances. It's even been used to help resolve the ambiguities in spouse selection!

On a more practical level, fuzzy decision-makers have been used to optimize the purchase of cars and VCRs. The Fuji Bank has developed a fuzzy decision-support system for securities trading.

FUZZINESS AND ASIAN NATIONS

If the names Nissan, Matsushita, and Fuji Bank jumped out at you, there's a reason. As they indicate, Japan is the world's leading producer of fuzzy-based commercial applications. Japanese scientists and engineers were among the earliest supporters of Lotfi Zadeh's work and, by the late 1960s, had introduced fuzziness in that country. In addition, research on fuzzy concepts and products is enthusiastically pursued in China. According to one survey, there are more fuzzy-oriented scientists and engineers there than in any other country.

Why has fuzzy logic caught on so easily in Asian nations, while struggling for commercial success in the United States and elsewhere in the West? There are two possibilities.

One answer is found in the different traditional cultures. As you saw earlier, one of the hallmarks of Western culture is the Aristotelian either–or approach to thought and action. Individual competitiveness and a separation of human actions from the forces of nature have helped foster the early development of technology in Europe and the United States.

The culture of China and Japan developed with different priorities. Strength and success were accomplished through consensus and accommodation among groups. This traditional attitude, so perplexing to Americans, is basic to Japanese business transactions today, from the smallest firm to the largest high-tech company. In addition, the forces of nature were traditionally expected to be balanced between complementary extremes—the Yin–Yang of Zen is an example. Fuzzy logic is much more compatible with these tenets than with the mathematically oriented Western concepts.

Or it may be that the research-oriented government–industry establishment in Japan is simply more open to new ideas and approaches than in management- and bottom line-oriented Western firms.

FUZZY SYSTEMS AND UNCERTAINTY

Two broad categories of uncertainty methods are currently in use—probabilistic and nonprobabilistic. Probabilistic and statistical techniques are generally applied throughout the natural and social sciences and are used extensively in artificial intelligence. Several nonprobabilistic methods have been devised for problem solving, particularly "intelligent," computerized solutions to real-world problems. In addition to fuzzy logic, they include default logic, the Dempster–Shafer theory of evidence, endorsement-based systems, and qualitative reasoning.

E-MAIL   These other methods of dealing with uncertainty provide in-
FROM   teresting context. But you don't have to understand them
DR. FUZZY   thoroughly to understand fuzziness.

Probability and Bayesian Methods

Probability theory is a formal examination of the likelihood (chance) that an event will occur, measured in terms of the ratio of the number of expected occurrences to the total number of possible occurrences. Probabilistic or stochastic methods describe a process in which imprecise or random events affect the values of variables, so that results can be given only in terms of probabilities.

For example, if you flip a normal coin, you have a 50–50 chance that it will come up heads. This is also the basis for various games of chance, such as craps (involving two six-sided dice) and the card game 21 or blackjack. On a more scholarly level, it's used in computerized Monte Carlo methods.

Bayes's rule or Bayesian decision theory is a widely used variation of probability theory that analyzes past uncertain situations and determines the probability that a certain event caused the known outcome. This analysis is then used to predict future outcomes. An example is predicting the accuracy of medical diagnosis, the causes of a group of symptoms, based on past experience. The rule itself was developed in the mid-eighteenth century by Thomas Bayes, but not popularized until the 1960s. It works best when large amounts of data are available.

Bayes's rule considers the probability of two future events both happening. Then, supposing that the first event occurs, takes the ratio of the probabilities of the two events as the probability of both occurring. In other words, the greater the confidence in the truth about a past fact or future occurrence, the more likely the fact is to be true or the event to occur.

Nonprobabilistic Methods

In addition to fuzzy logic, several extensions of crisp logic have been developed to deal with uncertainty.

Default Logic

In this system, the only true statements are the ones that contain what is known about the world (context or area of interest). This includes many commonsense assumptions and beliefs. For example, assume that traffic keeps to the right unless otherwise proven. This is the logic behind the computer language Prolog.

Default logic also lets the user add new statements as more knowledge is obtained, as long as they're based on previously accepted statements. For example, a system reasoning about the planet Mars might include the belief that it has no life, even though there's no direct proof.

Default reasoning and logic were developed by the Canadian Raymond Reiter in the late 1970s.

The Dempster–Shafer Theory of Evidence

The theory of evidence involves determining the weight of evidence and assigning degrees of belief to statements based on them. It was developed by the Americans Arthur Dempster in the 1960s and Glenn Shafer in the 1970s. But it's a generalization of a theory proposed by Johann Heinrich Lambert in 1764. For a given situation, the theory takes various bodies of evidence, uses a rule of combination that computes the sum of several belief functions, and creates a new belief function. The method can be adapted to fuzziness.

Endorsement

Endorsement involves identifying and naming the factors of certainty and uncertainty to justify beliefs and disbeliefs. The method, invented by the American Paul Cohen in the early 1980s, allows nonmathematical prioritizing of alternatives according to how likely each one is to succeed or how suitable it is for use. It also specifies how the sources interact and gives rules for ranking combinations of sources. For example, they can be sorted into likely and unlikely alternatives. Useful, for example, in prioritizing tasks by suitability or by likelihood of succeeding.

Endorsements are objects representing specific reasons for believing (positive endorsement) and disbelieving (negative endorsements) their associated evidence, which consists of logical propositions. Endorsement is the process of identifying factors related to certainty in a given situation. For example, in predicting tomorrow's weather, the conclusion that the weather is going to be fair, based on satellite weather pictures, is probably better endorsed than the conclusion that it is going to rain tomorrow, because that's when the Weather Service is having its office picnic.

Qualitative Reasoning

Qualitative reasoning is another commonsense-based method of deep reasoning about uncertainty that uses mainly linguistic, as well as numerical, data models to describe a problem and predict behavior. Qualitative reasoning has been used to study problems in physics, engineering, medicine, and computer science.

FUZZY SYSTEMS AND NEURAL NETWORKS

Today, fuzzy logic is being incorporated into crisp systems and teamed with other advanced techniques, such as neural networks, to produce enhanced results with less effort.

A neural network, also called *parallel distributed processing*, is the type of information processing modeled on processing by the human brain. Neural networks are increasingly being teamed with fuzzy logic to perform more effectively than either format can alone.

A neural network is a single- or multilayer network of nodes (computational elements) and weighted links (arcs) used for pattern matching, classification, and other nonnumeric problems. A network achieves "intelligent" results through many parallel computations without employing rules or other logical structures.

As in the brain, many nodes or neurons receive signals, process them, and "fire" other neurons. Each node receives many signals and, after processing them, sends signals to many nodes. A network is "trained" to recognize a pattern by strengthening signals (adjusting arc weights) that most efficiently lead to the desired result and weakening incorrect or inefficient signals. The network "remembers" this pattern and uses it when processing new data. Most networks are software, though some hardware has been developed.

Researchers are using neural networks to produce fuzzy rules. For fuzzy control systems, neural networks are used to determine which of the rules are the most effective for the process involved. The networks can perform this task more quickly and efficiently than can an evaluation of the control system. And turning the tables, fuzzy techniques are being used to design neural networks.

E-MAIL FROM DR. FUZZY

Are neuro-fuzzy systems practical?
  In Germany, a home washing machine now on the market learns to base its water use on the habits of the householder. A fuzzy system controls the machine's action, and a neural network fine-tunes the fuzzy system to make it as efficient as possible.

As you've seen from this overview, three major constructions are used in creating fuzzy systems—logical rules, sets, and cognitive maps. You'll meet all of them in greater detail in Chapter 2.

CHAPTER 2

FUZZY NUMBERS AND LOGIC

Scene: a deli counter.

"I want a couple of pounds of sliced cheeses. Give me about a half-pound each of swiss, cheddar, smoked gouda, and provolone."

The clerk works at the machine for a while and comes back with four mounds. "I went a little overboard on the swiss. Is 9 oz. OK? There's 9 oz. of the cheddar too, and a tad under 8 oz. of the provolone. We only had about 7 oz. of the gouda. Is that close enough?"

"That's fine," the customer says.

Somewhere early in life, we all learned that

$$2 + 2 = 4$$

at least in school and cash transactions. With flash cards, Cuisenaire rods, or by rote, we also absorbed the messages that

$$2 - 2 = 0$$
$$2 \times 2 = 4$$
$$2 \, / \, 2 = 1$$

There's nothing wrong with these precise—or crisp—numerical values. But as the scene in the deli shows, they're not always necessary or appropriate. Sometimes fuzzy numbers are better. At the cheese counter, "about half a pound" turned out to be anywhere from 7 oz. to 9 oz. and the service was quicker than if the clerk had laboriously cut exactly 8 oz. of each type of cheese. With the gouda, in fact, exactly 8 oz. would have been impossible to produce. All in all, the customer ended up with "a couple of pounds," as planned.

In this chapter, you'll delve more deeply into fuzziness, beginning with some basic concepts. The first of these is fuzzy numbers and fuzzy arithmetic operations. You'll also learn the fine art of creating fuzzy sets and performing fuzzy logical operations on them. And you'll discover how fuzzy sets, fuzzy rules of inference, and fuzzy operations differ from crisp ones. Finally, you'll begin learning the use of As–Do and As–Then problem-solving rules (the fuzzy equivalents of If–Then rules).

As always, Dr. Fuzzy will be available with more information and encouragement.

E-MAIL   Why learn the inner workings of fuzzy sets and rules?
FROM   They're the power behind most fuzzy systems out here in
DR. FUZZY   the real world.

Throughout the chapter, you can make use of the doctor's own series of fuzzy calculators, contained on the disk that accompanies this book. Each calculator is fully operational. You can compute the examples in the book, use your own examples, or press the **e** button to automatically load random numbers. The Introduction to the book contains instructions for using the disk programs with Windows 3.1 or above. Portions of the text that are related to calculator operations are marked with Dr. Fuzzy's cartouche. The doctor also provides context-sensitive help on request from the calculator screen.

Figure 2.1: A crisp 8.

As they say in Dr. Fuzzy's family, you have to crawl before you can fly, so we're going to ease into the doctor's Fuzzy World Tour with some very elementary fuzzy arithmetic.

Fortunately, the doctor likes to make tracks on wheels. Open the first calculator, FuzNum Calc by clicking on the Trike icon, and let's get rolling.

FUZZY NUMBERS

Back at the deli, a crisp "half pound" (8 oz.) registers on the scale as shown in Figure 2.1. Deli's don't have fuzzy scales (the Dept. of Weights and Measures would frown). But if they did, "about a half pound" might register like the representation in Figure 2.2.

Now try your own hand at fuzzy arithmetic with FuzNum Calc.

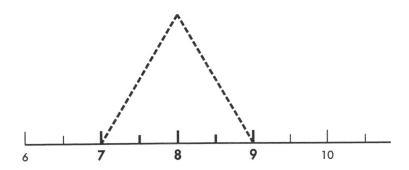

Figure 2.2: A fuzzy 8.

Meet FuzNum Calc

The fuzzy number calculator (Figure 2.3) has lots in common with the crisp calculator you probably have nearby. It has two Setup keys—Setup A and Setup B—that let you enter two numbers from the keypad. The minus (-) key allows negative numbers. Use the operation keys to perform addition (C=A+B), subtraction (C=A-B), multiplication (C=A×B), and division (C=A/B). It also has a Clear Entry (CE) key.

     The the numbers you enter, ranging from -9 to +9, appear on the calculator's screen. After you click the operation button, the screen displays the results on a scale from -100 to +100. The scale shifts automatically to display the numbers you enter and the results calculated. You can perform calculations on fuzzy numbers exclusively, crisp numbers exclusively, or

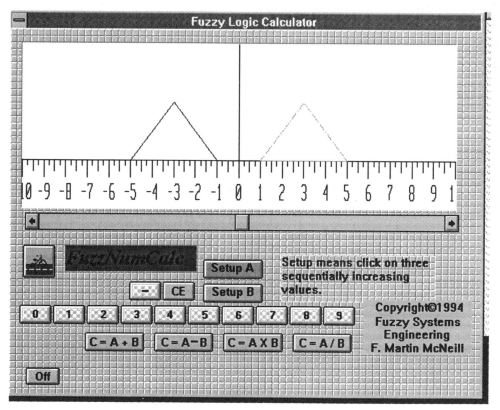

Figure 2.3: Opening screen of the FuzNum Calc.

combine fuzzy and crisp numbers in one operation. You can also move the scale yourself, using the slide bar just below the screen.

Performing Fuzzy Arithmetic

Each fuzzy number is represented by a triangle, with the apex above the number itself and the base extending across the numerical range of fuzziness. For instance—back to the cheese counter—fuzzy 8 rested on a base extending from 7 to 9.

Enter that fuzzy 8 into the calculator by clicking on the key labeled Setup A and clicking on the keypad numbers 7, 8, and 9. Positive numbers must be entered sequentially, from smallest to largest.

The triangle representing fuzzy 8 is shown in Figure 2.4. To think of the crisp number 8 in fuzzy terms, the range of the base is 8 and the apex is also 8. Enter it by clicking on Setup B and then clicking on the number 8 three times. The result is a vertical line superimposed on the fuzzy 8 (Figure 2.5).

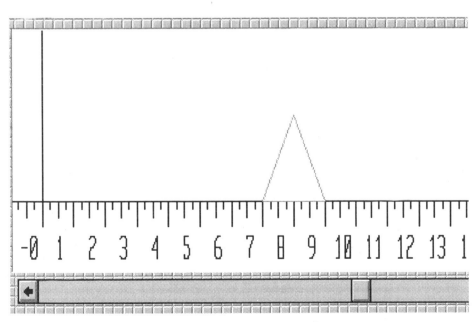

Figure 2.4: Fuzzy 8 triangle on the FuzNum Calc.

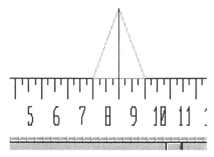

Figure 2.5: Crisp 8 and fuzzy 8 on the FuzNum Calc.

There's just one way you can represent any crisp number: crisp 8 is crisp 8. But a fuzzy number has any number of possible triangular shapes. The fuzzy number 8, with a base range of 7 to 9 forms an isosceles (symmetrical) triangle.

Try another triangular shape by clicking on Setup A and then the numbers 6, 8, and 9. This fuzzy number 8 has a different triangular representation—an asymmetric triangle (Figure 2.6).

For simplicity, FuzNum Calc presents the results as a symmetrical triangle. A more sophisticated computer would be able to represent results as asymmetrical triangles, as well.

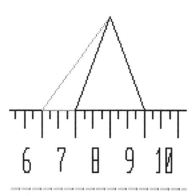

Figure 2.6: Two alternative fuzzy 8s on the FuzNum Calc.

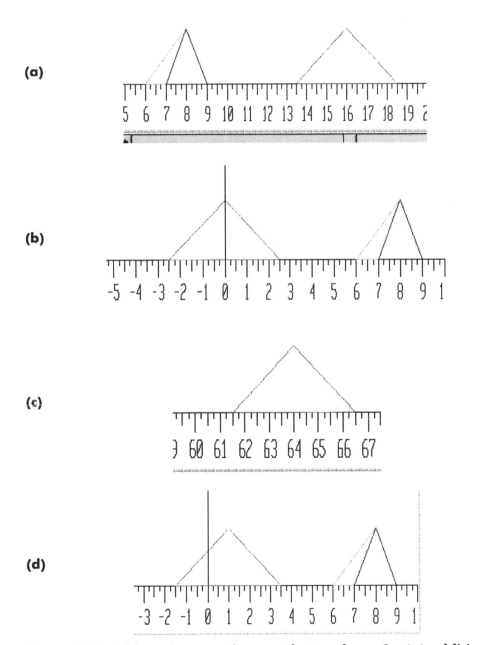

Figure 2.7: Arithmetic operations on the two fuzzy 8s: (a) addition, (b) subtraction, (c) multiplication, and (d) division.

TABLE 2.1: Crisp and fuzzy arithmetic operations

Crisp	Fuzzy
$a = 3$	$a = -2, 3, 8$
$b = 2$	$b = -1, 2, 7$
Addition: $a + b$	
$3 + 2 = 5$	$(-2, 3, 8) + (-1, 2, 7) = (-4, 5, 14)$
Subtraction: $a - b$	
$3 - 2 = 1$	$(-2, 3, 8) - (-1, 2, 7) = (-8, 1, 10)$
Multiplication: $a \times b$	
$3 \times 2 = 6$	$(-2, 3, 8) \times (-1, 2, 7) = (-3, 6, 15)$
Division: a / b	
$3 / 2 = 1.5$	$(-2, 3, 8) / (-1, 2, 7) = (-7.5, 1.5, 10.5)$

Now perform each of the four arithmetic functions on the two different fuzzy 8s by clicking on the appropriate operation button (Figure 2.7). Table 2.1 provides another set of examples to play with. Their results are even more dramatic.

E-MAIL FROM DR. FUZZY   Always enter numbers into FuzNum Calc from left to right as they appear on the scale.

Behind the Scenes With FuzNum Calc

Wonder how FuzNum Calc works? Here's Dr. Fuzzy's explanation. Each operation requires several steps, because the apex and the base are handled differently.

This example adds the fuzzy numbers (-1, 2, 5) and (3, 5, 7).

The actual arithmetic operations are performed only on the apex numbers, so that, for example, 2 + 5 = 7.

The base width is always handled the same way, regardless of the apex operation:

- The base ranges of the two fuzzy numbers are added together, forming the base of the arithmetic result. For instance, the base of fuzzy 2 ranges from -1 to +5,

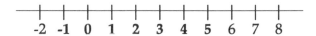

or 6. The base of fuzzy 2 ranges from +3 to +7,

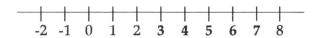

or 4. So 6 + 4 = 10.
- The sum is divided by 2. In the example, 10/2 yields a product of 5.
- Subtract this product from the result of the arithmetic operation on the apex number. For instance,

$$7 - 5 = 2$$

So 2 becomes the left-hand limit of the base.
- Add the product to the result of the arithmetic operation; for example,

$$7 + 5 = 12$$

making 12 the right-hand limit of the base.

The fuzzy result is (2, 7, 12).

Verify this by performing the operation on the fuzzy calculator. When you've finished with FuzNum Calc, press the OFF button to return to the main calculator menu. Once you've got fuzzy numbers cold, it's a short step to fuzzy sets.

**E-MAIL
FROM
DR. FUZZY**

You can turn off any of the fuzzy calculators by clicking on its OFF button.

FUZZY SETS

As the fuzzy calculator showed, any fuzzy number can be represented by a triangle. If you think of the calculator's linear scale as the horizontal line (abscissa) of a graph, you can easily convert the diagram to the repesentation of a fuzzy set by adding a vertical scale (Figure 2.8):

The values in this set—7, 8, and 9—have various degrees of membership in the set of Eightness. For instance, 7 and 9 have the least degree of membership, while 8 has the greatest degree of membership. You might represent these degrees of membership as shown in Table 2.2.

**E-MAIL
FROM
DR. FUZZY**

A triangular fuzzy set's apex has a membership value of 1. The base numbers have membership values of close to 0.

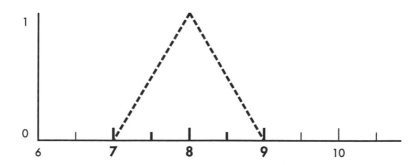

Figure 2.8: An example of fuzzy set of Eightness with a triangular membership function.

TABLE 2.2: The Set of "Eightness" with a Triangular Membership Function.

Member	Degree of Membership
7	.2
7.5	.8
8	1
8.5	.8
9	.2

The triangular membership function is the most frequently used function and the most practical, but other shapes are also used. One is the trapezoid, as shown in Figure 2.9. The trapezoid contains less information (fewer set points) than the triangle.

A fuzzy set can also be represented by a quadratic equation (involving squares, n^2, or numbers to the second power), which produces a continuous curve. Three shapes are possible, named for their appearance—the S function, the pi function, and the Z function (Figure 2.10).

Like other types of sets, fuzzy sets can be made to interact with each other to produce a usable result.

Most people have been exposed to classical set theory in school. In the world of fuzziness, classical set theory is called *crisp set theory*, in which set membership is limited to 0 or 1.

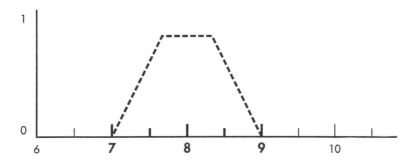

Figure 2.9: A fuzzy set of Eightness with a trapezoidal membership function.

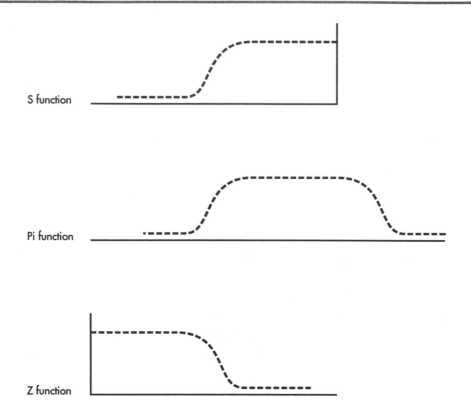

Figure 2.10: Graphs of the S function, the pi function, and the Z function.

Set Theory

The basic purpose of a set is to single out its elements from those in its domain or "universe of discourse" (Figure 2.11a). The relationship between two sets has two possibilities. Either they're partners merged in a larger entity or the relationship consists of the elements that they have in common.

Sets as partners (see Figure 2.11b) is called a *disjunction* (for single-element, or atomic, sets), using the symbol v, or a union (for multielement sets), using the symbol ∪. The disjunction or union of two sets means that any element belonging to either of the sets is included in the partnership. In the fuzzy world, this partnership expresses the maximum value for the two fuzzy sets involved.

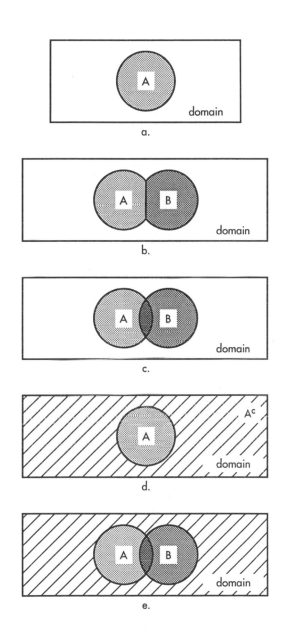

Figure 2.11: Crisp set operations: (a) Set A in a domain, (b) disjunction or union of Set A and Set B, (c) conjunction or intersection of Set A and Set B, (d) complement of Set A and Set Not-A in its domain, and (e) difference of Set A and Set B.

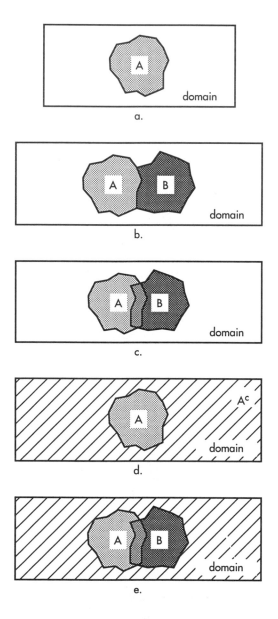

Figure 2.12: Fuzzy Set Operations: (a) fuzzy Set A in a domain, (b) disjunction or union (MAX) of fuzzy Set A and fuzzy Set B, (c) conjunction or intersection (MIN) of fuzzy Set A and fuzzy Set B, (d) complement of fuzzy Set A and Set Not-A in its domain, and (e) difference of fuzzy Set A and fuzzy Set B.

Set elements in common (Figure 2.11c) is called a conjunction (for single-element sets) or intersection ∩ (for multielement sets). A conjunction or intersection makes use of only those aspects of Set A and Set B that appear in both sets. In the fuzzy world, this partnership expresses the minimum value for the two fuzzy sets involved.

The part of the domain not in a set can also be characterized (Figure 2.11d)—what's called not-A (A^c). Not-A can also be written ~A or ≠A.

Set theory is closely linked to an operation in logic—the use of mathematics to find truth or correctness—called *implication*. (There's more on logical operations later in the chapter.) Implication is a statement that if the first of two expressions is true, then the second one is true also. For example, given the expressions A and B, if A is true, then B is also true. In other words,

A implies B

This can also be written

$$A \rightarrow B$$

As you've already experienced, fuzziness provides a great variety of ways for sets to interact—much more so than crispness. Looked at in this way, fuzzy sets are the more general way of approaching sets, and crisp sets are a special case of that generality. Figure 2.12 represents fuzzy versions of the principal set operations.

Set theory, fuzzy and crisp, can be better understood through use of another of the fuzzy calculators, the one named UniCalc. It calculates operations on single element sets. Change vehicles—or calculators—by clicking on the Bicycle icon to open UniCalc.

Touring UniCalc

UniCalc (Figure 2.13) provides a numeric/decimal keypad, the set operators conjunction (∧), disjunction (∨), not-A (~A), not-B (~B), and implication (the arrow key). To enter single-element values for Set A, click on the box by A and then on the desired keypad numbers. Follow the same procedure for Set B. You can enter any value between 0 and 1.

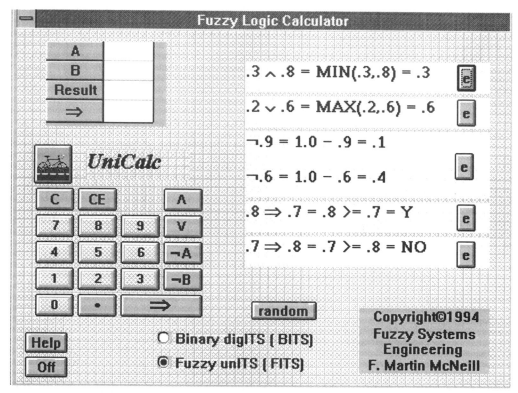

Figure 2.13: Opening screen of the UniCalc.

For example, click on the Set A box, then click on the value .3. Next, click on the B box and then on the value .8. Now click on the conjunction (∧) key. The Result box shows the calculation (Figure 2.14), here .3, representing the minimum of .8 and .3. Clicking on the disjunction (∨) key gives the result .8, the maximum value.

To see how the operations work for crisp sets, give set A the value 1 and set B the value 0. Then perform disjunction and conjunction (Figure 2.15).

To calculate complementation, enter a fuzzy value for set A, such as .7, and click on the ~A key. The value for ~A, which is .3 (1 − .7), appears in the A box (Figure 2.16).

You can demonstrate implication by entering values for A and B, then pressing the arrow key. (The implication method used here is the simplest: "contained within." There are many others.) If A implies B, YES appears in

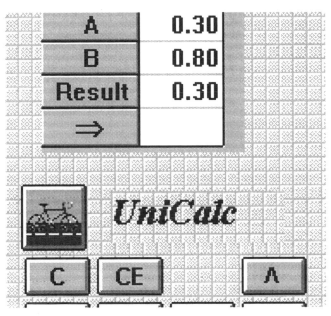

Figure 2.14: Fuzzy conjunction operation on UniCalc.

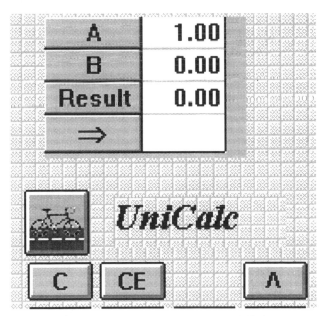

Figure 2.15: Crisp conjunction operation on UniCalc.

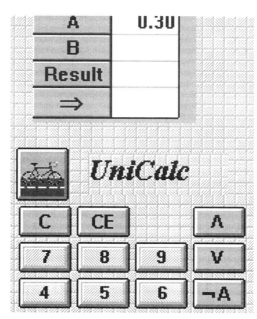

Figure 2.16: Complementation on UniCalc.

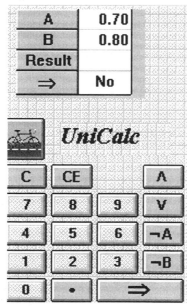

Figure 2.17: Implication on UniCalc.

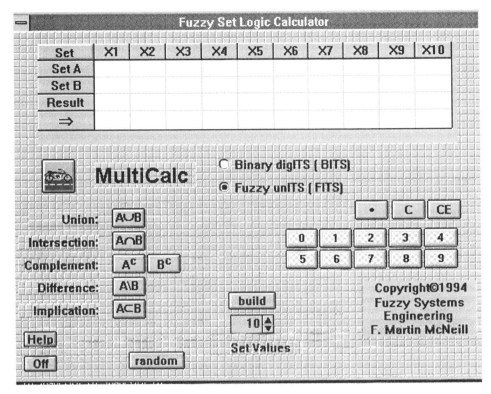

Figure 2.18: Opening screen of MultiCalc.

the Implication Box. If A doesn't imply B, NO appears in the Implication Box (Figure 2.17). For example, enter .7 as the A value and .6 as the B value. Since .7 is greater than .6, it implies .6, so the Implication Box displays YES. If you change B to .8 and click on the implication button, the Implication Box displays NO.

When you're finished with Unicalc, click on the OFF button. The next step in set theory involves sets with more than one element.

Multielement Sets

Now that you've warmed up by performing disjunction and conjunction on single element sets, Dr. Fuzzy will lead you to the next level of difficulty— multielement sets and additional set operations.

Click on the Convertible icon to open MultiCalc (Figure 2.18). Multi-Calc is an advanced version of UniCalc. It allows as many as 25 elements per set, with the comparable elements from each set calculated individually. MultiCalc also performs more operations. Three you've already experienced—union (A∪B), intersection (A∩B), and implication (A ⊂ B). It also calculates difference (A\B) and complement (A^c).

Select the number of set elements by clicking on the up and down arrows below the Build button to scroll through the numbers. Once the number you want appears in the window next to the arrows, click on the Build button. The number of elements you selected will be displayed at the top of the calculator.

If you choose more than 10 elements, you can display them by using the horizontal scroll bar.

**E-MAIL
FROM**   The number of elements in a set is called its *cardinality*.
DR. FUZZY

Select the number of set elements with the up and down arrows and Build. Next, enter values for the elements by clicking on each space, such as the Set A-X1 cell, then clicking on the desired values. For example, build three-element sets and enter the following values:

	X1	X2	X3	...	X25
Set A	.8	.2	.7		
Set B	1	.3	.4		
Result					

Union, Intersection, and Implication

Begin by reviewing the three fuzzy set operations you've already practiced—union (aka disjunction), intersection (conjunction), and implication. Before you actually use MultiCalc, take a mental self-test, *then* go electronic to see if you were successful.

When you click on the union operator

Union A ∪ B

the calculator responds with

Result   .1   .3   .7

Clicking on

Intersection A ∩ B

displays

Result   .8   .2   .4

Finally, click on

Implication A ⊂ B

The Implication row will display

No   No   Yes

The set values are now

Set A	.8	.2	.7
Set B	.1	.3	.4

Difference

Logical difference (A \ B) is set A minus the portion of it that is also in set B (see Figures 2.10e and 2.11e).

Clicking on

Difference A \ B

gives

$$\textbf{Result} \quad 0 \quad 0 \quad .3$$

If you click on the BITs button, the sets become crisp,

$$
\begin{array}{cccc}
\textbf{Set A} & 1 & 0 & 1 \\
\textbf{Set B} & 1 & 0 & 0 \\
\end{array}
$$

When you're finished with MultiCalc, click on the OFF button.

Complement

The set operation complement behaves differently in crisp and fuzzy sets. To explore it, Dr. Fuzzy provides CompCalc (Figure 2.19), which is very similar to

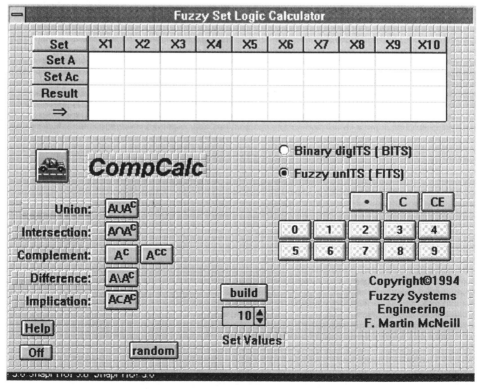

Figure 2.19: Opening screen of CompCalc.

MultiCalc, except that Set A^c replaces Set B. The same operations performed on MultiCalc's sets A and B can be performed on CompCalc's sets A and A^c. Open CompCalc by clicking on the Hatchback icon.

In the crisp world, the union

$$A \cup A^c$$

by definition includes the entire domain, as Figures 2.10b and 2.10d show.

The crisp intersection

$$A \cap A^c$$

is impossible, because the two are mutually exclusive (see Figures 2.11c and 2.11d).

In the "all or nothing" crisp world, $A \cup A^c$ is "all" and $A \cap A^c$ is "nothing." The fuzzy world presents other possibilities.

E-MAIL FROM DR. FUZZY Crisp $A \cup A^c$ is also known as the *law of excluded middle*. Crisp $A \cap A^c$ is also called the *law of contradiction*.

For starters, enter the fuzzy values

Set A   .8   .2   .7

Now click on

Complement A^c

This changes Set A to its complement

.2   .8   .3

and Set A^c, in turn, becomes

.8   .2   .7

Now perform fuzzy union by clicking on

Union A ∪ A^c

Rather than being the entire domain, as in crisp logic, the fuzzy union is the maximum for each pair of elements,

.8   .8   .7

When you perform fuzzy intersection by clicking on

Intersection A ∩ A^c

Rather than being mutually exclusive, as in crisp logic, the fuzzy intersection is the minimum for each pair of elements,

.2   .2   .3

This dramatic difference between crisp and fuzzy operations becomes even more vivid in the next section, on fuzzy logical rules.

CRISP AND FUZZY LOGIC

Set theory is closely related to the truth-finding logical statements called the *rules of inference*. As with sets, fuzzy rules of inference were devised a few decades ago, based on the much older crisp rules. And as with sets, the fuzzy rules are generalizations and the crisp rules are a special case within them.

Fuzzy logic shows that truth itself is fuzzy.

Rules of Inference

Rules of inference are rules for deriving *truths* from stated or proven truths. You've already met one of these rules disguised as the set operation called *implication*, in the form

$$A \to B$$

In logic, the same rule goes by the Latin name *modus ponens*, meaning affirmative mode, stated:

Given that A is true and A implies B, then B is also true.

This means that A implies B (or B is inferred from A), but B does not necessarily imply A. Modus ponens may also be stated in the form If-Then:

If A is true, *Then* B is also true

Crisp modus ponens can also be written

If   A
And $A \to B$
Then   B

A related rule, called *modus tollens*, meaning denial mode, can be written several ways:

Given that B is false and A implies B, then A is also false.

or

If   B
And $A \to B$
Then   A

Another way to present modus tollens is

$$A \to B \text{ means } \overline{B} \to \overline{A}$$

which is also called *contraposition*.

Here's how these two rules work together. For example, according to modus ponens,

> *If*   the apple is red
> *And*   a red apple is a ripe apple
> *Then* the apple is ripe

Modus tollens states,

> *If*   the apple is not ripe
> *And*   a red apple is a ripe apple
> *Then* the apple is not red

As a crisp situation, apple ripeness is simple to state. Either the apple is red and therefore ripe, or it's not ripe and therefore not red. Unfortunately, redness can be interpreted many ways. It includes many shades of color and an apple may be partly red. In the real world, an apple's redness and its ripeness constitute—you guessed it—a fuzzy situation.

Fortunately, a *generalized modus ponens* exists to handle the logic of fuzzy situations.

> *As*   the apple is very red
> *And*   a red apple is a ripe apple
> *Then* the apple is very ripe

A second fuzzy rule, called the *compositional rule of inference*, involves an explicit relationship:

> *As*   Apple #1 is very ripe
> *And*   Apple #2 is not quite as ripe as Apple #1
> *Then* Apple #2 is more or less ripe

Logical Statements

The set operators union and intersection also have counterparts in crisp logic. The most common way of representing them is with quantifiers in a type of statement structure called *predicate calculus*. The *or* of a union is represented by an *existential quantifier*, using the symbol ∃, read as "there exists." It states that there is at least one instance in which the statement is true. For example,

$$(\exists x) \, [\text{ripe (apple)}]$$

translated as "there exists one example of a ripe apple."

The *and* of an intersection is represented by a *universal quantifier*, using the symbol ⊃, read as "for all." It states that the statement is true in all instances, such as

$$(\supset x)\ [\text{apple}\ (x) \rightarrow \text{ripeness}\ (x)]$$

meaning all apples are ripe.

Is either statement logically true? It depends on the domain involved—the bowl on the table, the entire earth, or whatever.

Fuzzy logic encompasses "there exists" and "for all" and also provides intermediate statements between the two extremes.

The fuzzy logician, R. R. Yager, has shown that the word *few* is a less extreme form of *or* ("there exists"). Where the crisp statement says that a single instance of a ripe apple exists in the domain, *few* means that it might be Apple #1 *or* Apple #2 *or* . . .

The word *most* is a less extreme form of "for all." Rather than stating that all apples are ripe, *most* means that Apple #1 *and* Apple #2 *and* Apple #3 . . . in the domain are ripe.

The As–Then format is so handy in fuzzy thinking that it's used in the sets of word-based rules that control fuzzy systems.

AS–THEN AND AS–DO RULES—A SNEAK PREVIEW

Traditional or crisp rules are expressed in precise terms, such as:

> *If* the room temperature is less than 62 degrees,
> *Then* set the thermostat for 68 degrees

Even though most home thermostats are marked in degrees, that's not the way most people use them. "Turn up the heat a little," someone will say. Or "nudge the thermostat." In other words, home heating is really a fuzzy situation.

Fuzzy logic also uses If–Then-style rules, expressed by the form *As–Then* (the general form) or *As–Do* (the control form), instead. A fuzzy thermostat rule might read:

As the room temperature is cool,
Do turn on the heater to High

There could be fuzzy rules for parallel parking a car, the fuzzy situation first explored in Chapter 1. A flowchart of parallel parking might look like the one in Figure 2.20. The fuzzy rules might be

As your car is lined up next to the one in front of your space,
Then angle the car back into the space.

As you're approaching the curb,
Then turn the steering wheel slightly to adjust your angle.

As you're quite close to the curb,
Then turn the wheel so you can back up straight.

To recap set representations:

- Sets as partners

    disjunction ∨
    union ∪
    or
    maximum (MAX)
    existential quantifier ("there exists") ∃

**E-MAIL
FROM
DR. FUZZY**

- Fuzzy in-between quantifiers (examples)

    most
    few

- Set elements in common

    conjunction ∧
    intersection ∩
    and
    minimum (MIN)
    universal quantifier ("for all") ∀

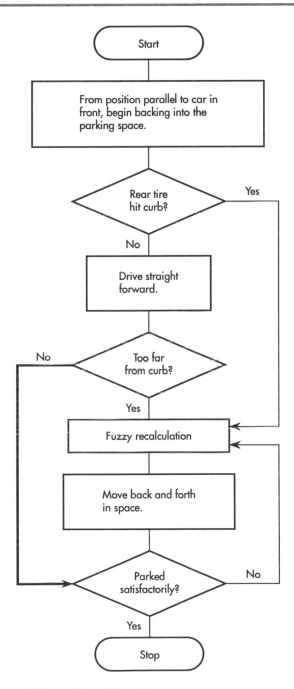

Figure 2.20: Parallel parking flowchart.

As you back up and the car doesn't move,
And as the rear tire's wedged against the curb,
Then go forward slowly, steering toward the curb until the
 rear tire straightens out.

As you're too far from the curb,
Then drive back and forth again, using shallower angles.

As you're close enough to the curb,
Then drive straight forward.

As you're a little too close to the car ahead,
Then back up a few inches.

As you've thunked into the bumper of the car in back,
Then drive forward a couple of inches.

As the car is positioned OK,
Then turn off the engine.

The use of fuzzy rules in a variety of real-world systems will be explored more deeply beginning in Chapter 3.

QUANTIFYING WORD-BASED RULES

Hedges—words that modify existing rules—played a large role in fuzzy programming in the olden days, when computers were slower and memory was scarcer than in the mid-1990s. Hedges include quantifiers such as *more or less, almost, higher than, often,* and *roughly.*

Today it's easier to write a new set of rules from scratch. Still, the concept of hedges remains available as a tool. And they're helpful in explaining the relationship between fuzzy words and crisp arithmetic. Fuzzy systems use what's called *fuzzification* (changing input values into fuzzy terms) and *defuzzification* (changing fuzzy output back into numerical values for system action).

TextCalc is Dr. Fuzzy's way for you to get acquainted with the process, using If–Then rules.

Open TextCalc (Figure 2.21) by clicking on the Van icon. The If–Then rule can accommodate as many as four *and*s, though you don't have to use all of them. The doctor provides several ways to modify this basic conjunction. You can change any *and* to an *or*, allowing disjunction. You can add negation (*not*). The hedges are *very* (quantified as the square of the original value) and *slightly* (the square root of the original value.) *Crisp* is also available as a hedge.

You can enter any value between 0 and 1. Value' displays the result of a hedge applied to that Value.

You can change a set name to any eight-letter phrase.

If you prefer As–Do to If–Then, clicking on the Alternate Prepositions button will make the change. In this calculator, clicking on the implication

---

Fuzzy Logic Calculator

Logic	Neg.	Hedge	Set		Value	Hedged
IF			Event A	IS	0.00	0.00
AND			Event B	IS	0.00	0.00
AND			Event C	IS	0.00	0.00
AND			Event D	IS	0.00	0.00
AND			Event E	IS	0.00	0.00
THEN			Result	IS	0.00	0.00

TextCalc

○ Binary digITS (BITS)
○ Fuzzy unITS (FITS)

| • | C | CE | ⇒ |

| 0 | 1 | 2 | 3 | 4 | 5 | 6 | 7 | 8 | 9 |

Help

random

Off

☐ Alternate Prepositions

Copyright©1994
Fuzzy Systems
Engineering
F. Martin McNeill

Figure 2.21: Opening screen of TextCalc.

**E-MAIL
FROM
DR. FUZZY**

In TextCalc,

$$VERY = VALUE^2$$
$$SLIGHTLY = \sqrt{VALUE}$$

(arrow) key performs the operation. Use the Clear button to remove all values.

Start with a simple If–And–Then set operation—the equivalent of a series of conjunctions—entering a value of .7 for Event A, .2 for Event B, .3 for Event C, .1 for Event D, and .6 for Event E. Now click on the arrow key. The Result Value is displayed as .1 (Figure 2.22).

Now double-click on Event A's Negation column, displaying *not* and recalculate. This time, the Value of Event A changes to .3. The Result remains .1, and the minimum value is .5 (Figure 2.23).

Again double-click on Event A's Negation cell to remove the *not*. Now double-click on Event D's Hedge column until *greatly* appears, and click on the arrow. The hedged value—the square of the Value—appears in the Value' column and a new Result is displayed (Figure 2.24).

Logic	Neg.	Hedge	Set		Value	Hedged
IF			Event A	IS	0.70	0.70
AND			Event B	IS	0.20	0.20
AND			Event C	IS	0.30	0.30
AND			Event D	IS	0.10	0.10
AND			Event E	IS	0.60	0.60
THEN			Result	IS	0.10	0.10

Figure 2.22: Simple *and* rule operation on TextCalc.

Logic	Neg.	Hedge	Set		Value	Hedged
IF	NOT		Event A	IS	0.70	0.30
AND			Event B	IS	0.20	0.20
AND			Event C	IS	0.30	0.30
AND			Event D	IS	0.10	0.10
AND			Event E	IS	0.60	0.60
THEN			Result	IS	0.10	0.10

Figure 2.23: Negation operation on TextCalc.

Logic	Neg.	Hedge	Set		Value	Hedged
IF			Event A	IS	0.70	0.70
AND			Event B	IS	0.20	0.20
AND			Event C	IS	0.30	0.30
AND		GREATLY	Event D	IS	0.10	0.01
AND			Event E	IS	0.60	0.60
THEN			Result	IS	0.01	0.01

Figure 2.24: TextCalc operation using the hedge _very_.

Double-click on the Hedge column again until _slightly_ is displayed and Recalculate, to see the effect of the value's square root.

To perform a disjunction, click on each Event's _and_, changing it to _or_, remove the hedge, and recalculate. Figure 2.25 shows the result.

To see the difference between fuzzy and crisp operations, first click on C to remove all the values, then double-click on each Event's Hedge column until _crisp_ is displayed. Now set the Value of Event A at 1, the Value of Event B at 0, and the others at either 0 or 1, and perform the various operations again.

Logic	Neg.	Hedge	Set		Value	Hedged
IF			Event A	IS	0.70	0.70
OR			Event B	IS	0.20	0.20
OR			Event C	IS	0.30	0.30
OR			Event D	IS	0.10	0.10
OR			Event E	IS	0.60	0.60
THEN			Result	IS	0.70	0.70

Figure 2.25: Simple *or* rule operation on TextCalc.

Dr. Fuzzy has now provided the means for hands-on experience with most of the basics of fuzziness. Chapter 3 begins an exploration of their use in real-world systems. Don't be surprised if the doctor shows up with some timely assistance.

CHAPTER 3

FUZZY SYSTEMS ON THE JOB

Today practical fuzzy systems are on the job in consumer products (washing machines, electric razors), industrial controllers (elevators), big public systems (a municipal subway), medical devices (cardiac pacemakers), and the business world (bond-rating systems). All these systems are solving some kind of problem, whether it's analyzing the past or predicting the future.

To understand better what fuzzy systems can do, Dr. Fuzzy says it's useful to take a look at the kinds of problem solving they're best at in the real world. If it's good enough for Dr. Fuzzy, it's gospel to us, so that'll be the first order of business in this chapter.

Next on the agenda is to take out the good doctor's fuzzy system blueprint and (verbally) construct one from the ground up, with the help of some of the doctor's nifty visual aids. Along the way, Dr. F. will point out some of the existing systems and reveal some of their inner secrets.

All this will be in preparation for the construction of a real computerized system—to come in Chapter 4.

FUZZY TOOLS

There are three basic types of fuzzy tools for problem solving. Almost all commercial fuzzy problem solvers are expert systems, descended from the control model developed by Ebrahim Mamdani. Another tool makes decisions—a model developed by Michael O'Hagan (Fuzzy Logic, Inc.). The final fuzzy tool, which describes how complex systems work, is called a *fuzzy cognitive map*, developed by Bart Kosko (University of Southern California). In this book you'll learn the anatomy and behavior of all three of these types of fuzzy systems, using modified versions of commercial software (on the accompanying disk).

Fuzzy Knowledge Builder™ for a Fuzzy Expert System

A *fuzzy expert system* (or *fuzzy knowledge-based system*) is a rule-based system composed of two modules—a *knowledge base*, mostly As–Then or As–Do rules, and an *inference engine*, which makes the rules work in response to system inputs.

The Fuzzy Knowledge Builder™ is a tool for creating the knowledge base. The job requires a partnership of two specialties, one for designing fuzzy systems and the other for expertise in the domain in question, such as engines, manufacturing processes, and other control systems. Almost all commercial fuzzy systems today are used for control.

One of the primary tasks of the designer is to learn how the expert works in the domain. For instance, how does an engine operator control the engine so it runs at maximum efficiency? Such learning isn't as easy as you might think.

True expertise is used intuitively, rather than thought through step by step. For instance, when you become an expert car driver or tennis player, you perform without consciously thinking about it.

When the designer does a good job, the computerized fuzzy knowledge-based system contains such intuitive expertise, then uses it to control a machine as close as possible to the way the human expert would.

A designer can use the Fuzzy Knowledge Builder™ to capture such expertise and put it in a knowledge base. It can be used in domains where the range of inputs and outputs are known ahead of time and don't change.

Its user is likely to be an engineer or designer. Later in this chapter you'll see how a simple fuzzy expert system is designed. You'll work with the Fuzzy Knowledge Builder™ in Chapter 4.

Fuzzy Decision-Maker™

The Fuzzy Decision-Maker™, which you'll meet in Chapter 5, is just that—a way to decide something, in business or in personal life, for example. It works on problems where the inputs are known and limited, and makes the best decision possible under the circumstances.

Fuzzy Thought Amplifier™

The Fuzzy Thought Amplifier™ (Chapter 6) is used to describe complex dynamic systems and it's intensely feedback driven. It's purpose is to model complex scenarios with more real-world confusion than other kinds of models can handle. For instance, it can be used to show the contributions of interacting conditions to a political or social system—such as apartheid in South Africa, a city's public health system, or war in the Middle East—and how changes can lead to stabilization or destabilization. A social scientist will find this tool valuable.

In a way, though, we all have to be engineers, social scientists, and business people. For example, the same problem situation can be examined differently with each tool.

FUZZY SYSTEMS

Most commercial fuzzy products are rule-based systems that control the operation of a mechanical or other device. The fuzzy controller receives current information fed back from the device as it operates. As Figure 3.1 shows, crisp information from the device is converted into fuzzy values that are processed by the fuzzy knowledge base. The fuzzy output is defuzzified (converted to crisp values) that change the device's operating conditions, such as slowing down motor speed or reducing operating temperature.

Business and management experts divide problems and problem-solving into several categories:

Prescriptive

Prescriptive problems require a specific decision. For example, a fast-food restaurant owner might need to find out how many customers she has at different times of day. With this information, she can determine how many employees she needs on duty at different times. This type of problem can be solved with the Fuzzy Decision Maker.

Descriptive

Here the need is to identify the problem. For instance, the fast-food restaurant owner may want to understand why customers have to stand in long lines at lunchtime. By describing how work is done in the restaurant, she may determine that the bottleneck is at the sandwich assembly station.

E-MAIL FROM DR. FUZZY   Electrical engineers will recognize this as a problem in queueing theory, in which plant identification describes the model. In queueing theory, one rule of thumb is that if the system is operating at 50% of capacity, it will cease to function effectively and become chaotic. Hungry burger lovers will know the feeling!

    This problem is an early phase of decision-making, so the Fuzzy Decision Maker™ will be useful for dealing with it and for the rest of the solution—how to deal with the bottleneck.

Optimizer

An optimizer establishes performance criteria, such as how many customers should be served per hour. It identifies the conditions or actions that allow the system to meet the criteria. Because it requires expert knowledge, it's a problem for the Fuzzy Knowledge Builder.™

Satisficing

A satisficing problem solver determines how to be "least worst"—how to maximize operations within already-established restrictions. The restaurant owner, for instance, might need to determine the maximum number of customers that can be served per hour, given a specified number of employees and the maximum number of burgers that can be cooked at once.

This type of problem would be suited to either the Fuzzy Decision Maker or the Fuzzy Knowledge Builder.™

E-MAIL FROM DR. FUZZY

Predictive

A predictive problem solver uses past results and projects them into the future (extrapolation). For example, the restaurant owner may analyze how many customers ate at the restaurant on the day after Thanksgiving last year, then use that information to predict the crowd on that same day this year. Predictive problems can be solved with the Fuzzy Knowledge Builder™ or the Fuzzy Thought Amplifier™.

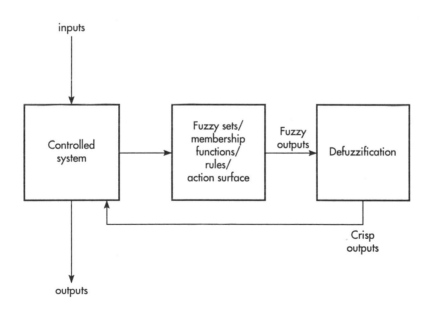

Figure 3.1: Diagram of a fuzzy controller.

A modern fuzzy control system isn't much different from the first one devised by Ebrahim Mamdani in the 1970s, an automated control system for the speed of a steam engine. Mamdani's fuzzy system had two inputs and two outputs and incorporated the expertise of a human machine operator with a set of fuzzy rules. The fuzzy system received two inputs from the steam engine as it operated, the engine speed and boiler pressure. It processed the information through the knowledge base and produced two outputs, the degree of throttle (the valve that controls how much steam enters the engine) and the boiler temperature.

CREATING A FUZZY CONTROL SYSTEM

The standard method of creating a fuzzy control system involves identifying and naming the fuzzy inputs and outputs, creating the fuzzy membership function for each, constructing the rule base, and deciding how the action will be carried out.

The early parts of any fuzzy control design are drawn from the intuitive experience of an expert. For instance, suppose you want to create a fuzzy system that uses the way a bicycle rider determines how and when to put on the brake for a stop sign. Dr. Fuzzy has found a bike-riding expert (who modestly wants to be known only as BikeRider) for a down-to-the-basics system with two inputs and one output. Dr. Fuzzy calls this system BikeBraker.

Identify and Name Fuzzy Inputs

As any bike rider might guess, the two inputs are *speed* and *distance*. The next step is to identify the fuzzy ranges of each.

Speed

BikeRider, a city dweller, pedals along streets posted from 35 mph to 50 mph. The particular block in question is posted at 35 mph. Naturally, the BikeRider is a law-abiding citizen, but occasionally gets carried away, and so identifies four fuzzy speed ranges: *Stopped, Slow, Pretty Fast,* and *Real Fast* (see Table 3.1).

TABLE 3.1: Names and Ranges for BikeBraker Speed Input.

Name	Range (mph)
Stopped	0–2
Slow	1–10
Pretty Fast	5–30
Real Fast	25–50

Distance

Let's say that a city block is about 660 feet long, and that all consideration of braking comes within a quarter-block (about 165 feet) of the stop sign. BikeRider gives the system five fuzzy distance from stop sign ranges: *At, Real Close, Near, Pretty Far,* and *Real Far* (see Table 3.2).

In most control systems, the majority of the action is in the lower ranges. This is the case in the BikeRider's fuzzy speed and distance.

TABLE 3.2: Names and Ranges for BikeBraker Distance Input.

Name	Range (feet)
At	0–5
Real Close	0–40
Near	20–80
Pretty Far	60–120
Real Far	100–165

Identify and Name Fuzzy Output

There's just one output for BikeBraker, *Brake Pressure*, measured in percentage of total braking capacity—in this case the maximum squeeze of the calipers on the tire. BikeRider's fuzzy braking ranges are *None, Light, Medium,* and *Squeeze Hard* (see Table 3.3). As with the inputs, most control system outputs are in the lower ranges.

TABLE 3.3: Names and Ranges for BikeBraker Braking Output.

Name	Range (%)
None	0–1
Light	1–30
Medium	25–75
Squeeze Hard	65–100

Create the Fuzzy Membership Functions

Fuzzy control systems are "expert" systems, meaning they're modeled on the expert experience of real people. The next step is to incorporate such experience in defining the fuzzy membership functions for each input and output (Figure 3.2).

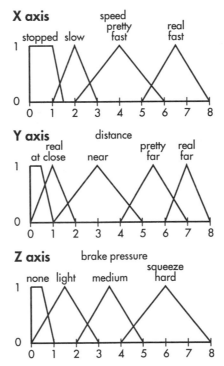

Figure 3.2: Fuzzy membership functions for inputs and output: (a) speed input, (b) distance input, and (c) braking output.

Construct the Rule Base

Now we write the rules that will translate the inputs into the actual outputs. The first thing to do is to make a matrix—a spreadsheet—of the inputs. The BikeBraker matrix places Speed along the horizontal and Distance along the vertical (Figure 3.3).

Designing the Interactions

The matrix provides one empty cell for each Distance–Speed combination. What goes in the cells? Each cell can contain a fuzzy output action, though they don't all need to be filled in. Dr. Fuzzy likes to be thorough, so the matrix is filled with an action or no action to each Speed–Distance combo (Figure 3.4).

The actions are those designed into the output membership functions (Figure 3.2). One purpose of the matrix is to look at the input-output

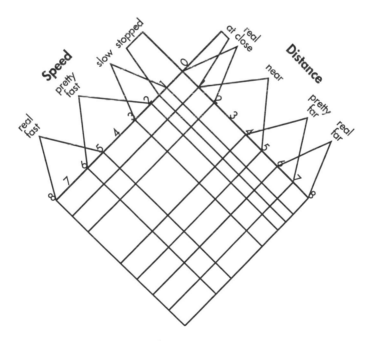

Figure 3.3: Input matrix.

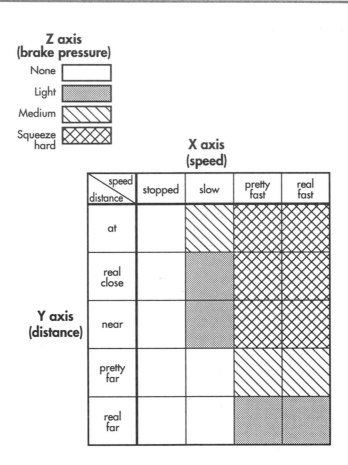

Figure 3.4: Matrix with actions.

relationship as a whole to see if it accounts for both normal operation and abnormal events. At this point, the control system designer can refine the inputs and outputs and add additional membership functions, if necessary.

Writing the Rules

Next, the system designer uses the matrix as the basis for the actual rules, one for each matrix cell. Table 3.4 displays the 20 As–Then rules, each of which is an *and* rule.

TABLE 3.4: BikeBraker Rules.

As you are *At* the stop sign *and* you are *Stopped*
Then brake pressure is *None*

As you are Real Close to the stop sign *and* you are Stopped
Then brake pressure is None

As you are Near the stop sign *and* you are Stopped
Then brake pressure is None

As you are Pretty Far from the stop sign *and* you are Stopped
Then brake pressure is None

As you are Real Far from the stop sign *and* you are Stopped
Then brake pressure is None

As you are At the stop sign *and* you are going Slow
Then brake pressure is Medium

As you are Real Close to the stop sign *and* you are going Slow
Then brake pressure is Light

As you are Near the stop sign *and* you are going Slow
Then brake pressure is Light

As you are Pretty Far from the stop sign *and* you are going Slow
Then brake pressure is None

As you are Real Far from the stop sign *and* you are going Slow
Then brake pressure is None

As you are At the stop sign *and* you are going Pretty Fast
Then brake pressure is Squeeze Hard

As you are Real Close to the stop sign *and* you are going Pretty Fast
Then brake pressure is Squeeze Hard

As you are Near the stop sign *and* you are going Pretty Fast
Then brake pressure is Squeeze Hard

As you are Pretty Far from the stop sign *and* you are going Pretty Fast
Then brake pressure is Medium

As you are Real Far from the stop sign *and* you are going Pretty Fast
Then brake pressure is Light

As you are At the stop sign *and* you are going Real Fast
Then brake pressure is Squeeze Hard

As you are Real Close to the stop sign *and* you are going Real Fast
Then brake pressure is Squeeze Hard   *(continued)*

TABLE 3.4: BikeBraker Rules *(continued)*.

As you are Near the stop sign *and* you are going Real Fast
Then brake pressure is Squeeze Hard

As you are Pretty Far from the stop sign *and* you are going Real Fast
Then brake pressure is Medium

As you are Real Far from the stop sign *and* you are going Real Fast
Then brake pressure is Light

Making the Rules "Work Fuzzy"

When you last saw the rules matrix (Figure 3.4), it didn't look very fuzzy. How does it become fuzzy? To find out, it's helpful to begin with a "what if": What if the rules and matrix were crisp?

Here's how the system would work. When BikeBraker operated, the combination of Speed and Distance would apply to just one matrix cell. The rule written in that cell would fire. The rest of the rules wouldn't fire.

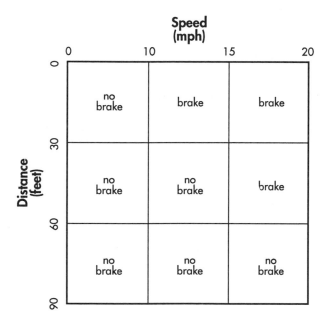

Figure 3.5: Example of a crisp rule matrix.

In other words, as you've seen with Dr. Fuzzy's calculators, the input membership function for each Speed and Distance value would be either 1 or 0 (Figure 3.5).

As the membership function for Speed is 1 (*as* Speed is 15, for instance)
And the membership function for Distance is 1 (*as* Distance is 20),
Then the rule in the matrix cell (15–20) will fire. No other rule fires.

Since BikeBraker is fuzzy, there are degrees of membership function, and any crisp input falls in several membership fuzzy sets. Also, fuzzy rules "fire fuzzy." The degree of membership determines the degree to which the rule fires. Because several sets are involved, several rules fire.

For instance, suppose BikeRider is going 7 mph and is about 25 feet from the stop sign. The speed of 7 mph falls into two fuzzy Speed sets,

Slow: 1–10 mph   and   Pretty Fast: 5–30 mph

The distance of 25 feet also falls into two fuzzy Distance sets,

Near: 20–80 feet   Real Close: 0 40 feet

This means that four matrix cells are involved will fire, as Figure 3.6 shows (x is a membership value between 0 and 1).

- *As* the membership function for SLOW Speed is x
  And the membership function for NEAR Distance is x,
  Then the rule in the Slow–Near matrix cell fires *partially*
- *As* the membership function for Slow Speed is x
  And the membership function for Real Close Distance is x,
  Then the rule in the Slow–Real Close matrix cell fires *partially*
- *As* the membership function for Pretty Fast Speed is x
  And the membership function for Near Distance is x,
  Then the rule in the Pretty Fast–Near matrix cell fires *partially*
- *As* the membership function for Pretty Fast Speed is x
  And the membership function for Real Close Distance is x,
  Then the rule in the Pretty Fast–Real Close matrix cell fires
    partially

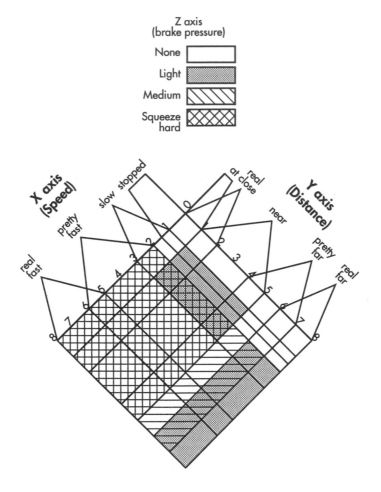

Figure 3.6: Example of fuzzy rules firing.

Decide How to Execute the Actions

How do you calculate the degree to which each rule fires? The first step is to ask another question: If the inputs are in the *vertical* and *horizontal* dimensions, what dimension is the output? One way to answer that question is to give the matrix a third dimension, *depth*. This is the dimension that turns a square into a cube. How do you create this third dimension?

Here's one way to really put your whole self into the project. Stand up and point your right foot at a 45° to the right and your left foot at a 45° to the left. You can think of your feet as standing on the vertical and horizontal dimensions. Your body is the depth dimension. You can label the horizontal dimension *x*, the vertical dimension *y*, and the depth dimension *z*.

E-MAIL FROM DR. FUZZY   Three-dimensional space is also known as *Cartesian space*, with each dimension called an *axis*—the *x*, *y*, and *z* axes.

Another way to think about the three dimensions is to imagine the process of setting up a tent with poles and canvas. First, place the poles

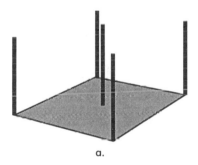

a.

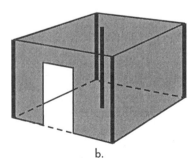

b.

Figure 3.7: Tent raising: (a) tent poles and 2-D canvas, and (b) 3-D tent.

upright (Figure 3.7a). Now raise the canvas from the ground (the x and y dimensions) so that some of it is supported by the tops of the poles (z dimension), as Figure 3.7b.

Suppose instead of a tent, you want to erect a canopy over a given space—for instance, to shade the table at a backyard picnic. You'll probably have four poles of the same length, one at each corner, with the canvas secured to the tops of the poles. If the ground is level, the canopy will be more or less level, too (Figure 3.8).

But suppose you're raising the canopy for another purpose and you don't want it to be level. Maybe you want it to represent something that's happening on the ground—some kind of an interaction between the vertical and horizontal dimensions. In this case, you might want to erect lots of different-length poles that are spaced evenly. Once you have the poles placed, you attach the canvas to the top of each pole. What does this canopy look like? Instead of being more or less flat, the canopy's surface looks hilly, with peaks and valleys determined by the heights of the poles (Figure 3.9).

In this scenario, the "ground" represents the fuzzy input matrix, the poles represent the interaction of the two inputs, and the canopy is the output or the degree to which the defuzzied rules fire. Dr. Fuzzy calls the hilly "canopy" a *fuzzy action surface*, because it's actually where the action is. The fuzzy action surface (also called a *fuzzy estimation surface*) performs the same action of reacting to the fuzzy inputs with appropriate outputs. In other words, it performs the action or judgment as the human expert under the same circumstances: It estimates, interpolates, extrapolates, transforms, decides, categorizes, extracts features, controls, simplifies, or anticipates!

E-MAIL FROM DR. FUZZY
The original way of creating a fuzzy action surface was with what's called a *fuzzy associative memory* (FAM). FAMs are math and theory intensive but not too practical. Somewhere along the way, someone discovered that the triangular membership functions do the same work and are much easier to work with.

Learn more about FAMs in Appendix A.

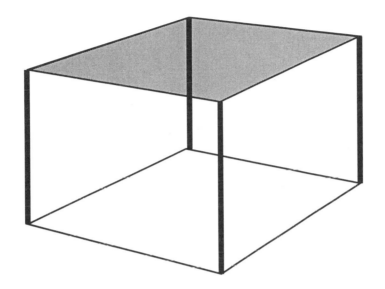

Figure 3.8: A level canopy.

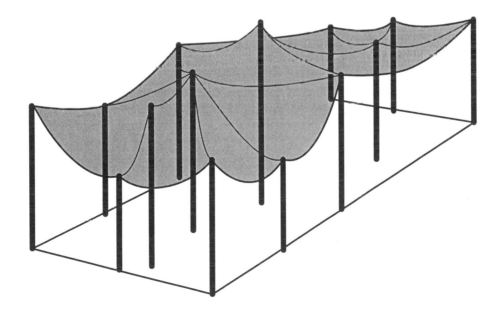

Figure 3.9: A "hilly" canopy.

How is the output determined? Fuzzy logic is applied to the fuzzy rules. For instance, suppose the first of the four (previous) matrix cells is given the following input values:

> *As* the membership function for Slow Speed is .7
> *And* the membership function for Near Distance is .5,

This is an *and* rule, meaning that the *then* portion is a conjunction or *minimum* of the input values:

$$.7 \wedge .5$$

or

$$.5$$

resulting in:

> *Then* the rule in the Slow–Near matrix cell fires at a strength of .5

Conjunctions are performed on each of the four cells so that, say, the *then* values are

$$.5 \quad .3 \quad .7 \quad .8$$

Two methods of translating the rule firings into crisp output values are the *fuzzy Or* and the *centroid*.

Fuzzy Or

To determine the output value, perform a disjunction (*or* or *maximum*) operation on them,

$$.5 \vee .3 \vee .7 \vee .8$$

resulting in

$$.8$$

This means that the crisp output value would be .8 of the maximum braking power.

Centroid

The centroid is the center of the output membership function adjusted to the degree of rule firing. It works like this.

- First, modify each affected output membership function so it's cut off at the strength indicated by the rule firing. For instance, if the rule involving a membership function fires at .5, the curve covers the membership function area from 0 to .5, rather than 0 to 1. The area between .5 and 1 is lopped off, turning a triangle into a trapezoid shape (Figure 3.10).
- Next, calculate the centroid—the center of activity—of each modified output membership curve.
- Take a weighted sum of the centroids; this becomes the crisp output number.

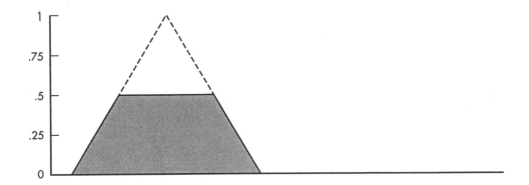

Figure 3.10: A triangular membership function modified to a trapezoid.

Here's a recap of the five-step process of creating a rule-based fuzzy system.

1. Identify the inputs and their ranges and name them.
2. Identify the output and their ranges and name them.
3. Create the degree of fuzzy membership function for each input and output.
4. Construct the rule base that the system will operate under.
5. Decide how the action will be executed by assigning strengths to the rules and defuzzification.

E-MAIL FROM DR. FUZZY

For the knowledge base, the expert defines the input and output observation (the descriptive words) and the range (the fuzzy number range). The expert also defines the consequent output for each input (the rule). The designer defines the membership functions for inputs and outputs.

The knowledge base is then put into action in an inference engine—a computer program that can take actual inputs, let them fire the rules, and export outputs to the domain system.

FUZZY BUSINESS SYSTEMS

Bond Rating

Hiroyuki Okada of Fujitsu Laboratories Ltd. and colleagues have developed a neuro-fuzzy system for rating the investment safety of bonds, a prescriptive problem. The research system works through the firing of a set of 10 fuzzy rules. The fuzzy membership functions or the rules are adjusted or modified in a neural network.

The basic rules cover the inputs and the financial condition of the company issuing the bonds: ordinary profit (with membership functions for large, medium, small), owned capital (large, small), interest coverage ratio (high, low), long-term loan ratio (low, high), and owned capital ratio (low). Outputs are high, medium, and low; they're based on the bond ratings according to the Japan Bond Research Institute—AAA (the highest rating), AA, A, BBB, BB, and B.

The system uses two classes of rules. The basic rules, which receive an initial weighting of 1, are related to ordinary profits. Auxiliary rules cover the other inputs and receive an initial weighting of 0.2. The weightings change after learning takes place in a multilayered neural network.

The basic rules are these:

1. *As* ordinary profit is large, *then* rating is high. After learning, the weight increased from 1 to about 2.5.
2. *As* ordinary profit is medium, *then* rating is medium. After learning, the weight decreased from 1 to about .7.
3. *As* ordinary profit is small, *then* rating is low. After learning, the weight decreased from 1 to about .6.

The auxiliary rules are these:

4. *As* owned capital is large, *then* rating is high. After learning, the weight increased from .2 to about .8.
5. *As* owned capital is small, *then* rating is low. After learning, the weight increased from .2 to about 2.2.
6. *As* interest coverage ratio is high, *then* rating is high. After learning, the weight increased from .2 to about .7.
7. *As* interest coverage ratio is low, *then* rating is low. After learning, the weight decreased from .2 to about −.2.
8. *As* long-term loan ratio is low, *then* rating is high. After learning, the weight decreased from .2 to about 0.
9. *As* long-term loan ratio is high, *then* rating is low. After learning, the weight decreased from .2 to about −.1.
10. *As* owned capital ratio is low, *then* rating is low. After learning, the weight increased from .2 to about 2.2.

Creditworthiness Based on Data Analysis

C. von Altrock and B. Krause, at the German software company INFORM, developed a method of data analysis that can be used for a variety of purposes—it's being used by Mercedes-Benz in the design of automobile and truck parts—and determining creditworthiness. Rather than MIN/MAX operators, the system uses an intermediate *gamma operator*.

The analysis for this optimizing problem takes a several-step reduction of multiple fuzzy inputs to a single input, which is evaluated for output. For instance, the initial creditworthiness inputs are Property and Other net property, which are reduced to Security; Income and Continuity (Liquidity); Potential and Motivation (Potential); and Economic Thinking and Conformity (Business Behavior). Next, the four intermediate variables are reduced: Security and Liquidity become Financial Basis; Potential and Business Behavior become Personality. Finally, Financial Basis and Personality are reduced to Creditworthiness.

INDUSTRIAL FUZZY SYSTEMS

Oil Recovery

An extremely complex fuzzy system was designed by W. J. Parkinson and K. H. Duerre at Los Alamos Scientific Laboratory as an experiment to determine the best techniques for improving the recovery of oil from the ground, an optimizing problem. The purpose is to extract more of the estimated two-thirds of known oil that cannot be removed with conventional pumping and other existing technology.

The four categories of alternative methods are uses of chemicals, such as polymers, surfactants (detergents), and alkalines; injection of gases, including hydrocarbons, carbon dioxide, and nitrogen and flue gas; heat, either combustion or steam flooding; and the use of microbes.

Input membership functions (such as preferred, fair, possible, poor) for various characteristics of the oil formation were written for each recovery method. The characteristics included gravity, viscosity, carbon composition, salinity, oil saturation, type of rock formation, thickness of the formation, average permeability, well depth, temperature, and porosity.

The single output was a score for the best recovery method, whose membership functions were not feasible, very poor, poor, possible, fair, good, medium good, and preferred. Crisp values were calculated by the centroid method.

FUZZY–NEURO SEWAGE PUMPING STATION

Chinese scientist Hong Chen and colleagues at Osaka Electro-Communication University have designed an automated combined storm–sewage pumping station for Shanghai, China. This optimizing probem solver uses fuzzy–neural network system to regulate six pumps—three that control the flow of sewage and three for storm runoff—so the combined flow doesn't overwhelm the treatment facility and let untreated sewage back up, an extremely nonlinear problem. The system has been run in simulation and may be installed in the Shanghai treatment plant.

The difficulty with crisp pump controllers is that they often start too late to prevent backflow, run too long, or stop too soon. The fuzzy system was designed to correct these deficiencies by determining when the pumps should be started and stopped for various conditions.

The fuzzy controller's two-dimensional rules matrix governs when the pumps are started and stopped. One dimension is the water level in an inlet storage well, defined in meters. Its fuzzy membership sets are very small, rather small, middle, rather large, and very large. The other dimension is the change in the water level, in meters per minute. Its sets are negative big, negative small, zero, positive small, and positive big. Membership functions are adjusted according to the weather—sunny, rainy, and stormy.

FUZZY INSULIN INFUSION SYSTEM FOR DIABETICS

Shigeru Kageyama and colleagues at Jikei University School of Medicine have developed an experimental fuzzy method that optimizes the timing and amount of insulin that diabetic patients receive through an insulin pump. People with insulin-dependent diabetes can't metabolize sugar successfully because their bodies produce too little or none of the hormone insulin.

Normally, when someone eats a meal, the body's level of glucose rises. To control this, the body also increases its insulin production. The most natural way to control insulin-dependent diabetics' deficiency is by infusion of insulin through a small pump system that's time released to imitate the natural method.

Conventional infusion methods are based on the level of glucose (simple sugar) in the blood. But this doesn't work as well as it should, because there's a time lag between eating a meal and the increase in the glucose level in the blood. This allows greater-than-normal swings in glucose levels in the patient's blood.

The fuzzy method takes the timing of the meal into account, as well as the blood glucose level. This way, the pump begins infusing insulin sooner than the standard pump method. As a result, the blood glucose level doesn't increase above that in nondiabetic people.

FUZZY CONSUMER PRODUCTS

Vacuum Cleaner

The American company NeuraLogix has drawn up a three input–four output fuzzy system for optimally controlling a vacuum cleaner. Inputs are vacuum pressure (very low, low, medium, high, very high), quantity of dirt (low, medium low, medium, medium high, high), texture of floor (smooth, medium, rough). Outputs are vacuum control, beater brush height, beater brush speed, cleanness indicator, and change bag indicator.

Washing Machine

Bert Hellenthal, from the German company INFORM, has developed a neuro–fuzzy washing machine that's now on sale in Europe. Based on the characteristics of the wash load during initial agitation, the system calculates the speed, water level, and time required for optimal execution of washing, the rinse cycle, and the number of rinse cycles.

The neural network is used to develop fuzzy rules and allow learning to take place. In addition, the user can override the system at any time, providing the system with interactive learning.

Fuzzy Air Conditioner

Mitsubishi Electric Corp. has developed a room air-conditioning system that's now on sale in Japan. An ordinary crisp thermostat takes a temperature setting, such as 68°, then keeps the temperature within several degrees of the setting. The cooling system is either ON or OFF. The thermostat is designed to keep the temperature with a range of 3° on either side of the setting of 68—between 65° and 68°.

Mitsubishi's fuzzy temperature controller allows any degree of operation from ON to OFF for optimal air-conditioning. It is said to improve the room's temperature consistency, it's three times as stable as a crisp system, and it provides a 24% power savings.

Every Japanese air conditioner contains a heat sensor, so can detect whether the room is occupied. It can then direct the air upward if people are present or downward if the room's empty. Mitsubishi's fuzzy air conditioner also can learn the room's characteristics and fine-tune its own operation.

CHAPTER 4

FUZZY KNOWLEDGE
BUILDER™

In Chapter 3 you became acquainted with the idea of a fuzzy expert system and its two parts, a knowledge base and an inference engine. In this chapter, you'll learn how to create practical, useful fuzzy knowledge bases, try them out in a simple inference engine, and learn how to format a knowledge base for use in several commercially available inference environments.

To construct a knowledge base, you'll use a special version of the commercial product, Fuzzy Knowledge Builder™. The commercial product has been used for projects ranging from improving the efficiency of a work vehicle to analyzing the quality of electrical power produced by an industrial supply system.

You'll "build along with Dr. Fuzzy," creating knowledge bases from two real-world scenarios—one for a graphics-display lunar lander and one for a personnel detection system. When the construction process is done, you can save each knowledge base in a format to run in an inference engine.

Dr. Fuzzy has provided three "plain wrap" inference engines—two written in QBASIC and one (provided by Motorola) in 68HC05 assembly language—along with test files for all of them. And you'll learn how to use the scenario's knowledge-base files in the QBASIC engines.

E-MAIL FROM DR. FUZZY

There are two schools of thought about the connection of a fuzzy expert system's knowledge base and its inference engine. Some developers like the knowledge base to be imbedded in the inference engine, so the code-writing is more efficient. A Fuzz-C (Bytecraft, Inc.) C language imbedded inference engine is included on the disk.

Others prefer to have the two separate for increased flexibility. For instance, the Fuzzy Knowledge Builder™ lets you develop the knowledge base independently, determine which inference engine it'll be used in, then save it in a format for the specific engine. A C language inference engine on the disk depicts this form. Also see Appendix D.

KNOWLEDGE BUILDER'S DESIGN

Creating a knowledge base with the Fuzzy Knowledge Builder™ follows the five-step design you are familiar with from Chapter 3:

1. Identify the inputs and their ranges and name them.
2. Identify the output and its ranges and name it.
3. Create the fuzzy membership function for each input and output.
4. Translate the interaction of the inputs and outputs into As–Then rules. If all are *and* rules, this interaction can be represented as a matrix. (The Fuzzy Knowledge Builder™ was the first product to use a graphical matrix representation of the rules.) If *and* and *or* rules are allowed, fewer rules are required, but the clarity of matrix representation is lost.
5. Decide on the inference engine that will act on the specific inputs and the knowledge base to produce the specific defuzzified output.

Program Organization

The program has six sections:

- A Naming menu for Steps 1–3,
- A rules Matrix Builder for Steps 4–5,
- Tools for fine-tuning the knowledge base, including a Set Shaper for adjusting one or more fuzzy membership functions, the ability to change one or more rules on the action matrix, a variety of ways to display the information on the screen, and a cellular automata tool for smoothing the matrix cell boundaries or interpolating between rules,
- A set of Viewers that let you examine the action surface from various angles—3-D Viewer, Gradient Viewer, and Profile Viewer,
- An Action Tester for trying out various inputs, observing the output, and deciding whether set or rule editing is required,
- A knowledge base File Generator, creating a file that's compatible with QBASIC, C, or Motorola 68HC05 assembly language.

E-MAIL FROM DR. FUZZY

The book version of the Fuzzy Knowledge Builder™ allows a maximum of two input dimensions and one output dimension. The commercial version allows additional dimensions for both input and output.

Program File Structure

The Fuzzy Knowledge Builder™ stores each knowledge-base project in two files. One, with the extension *.rul,* contains the rules. The other, with an *.fam* extension, contains everything else. For example, the knowledge base named TEST, which is on the disk, is stored in the files TEST.RUL and TEST.FAM. The .fam files are listed in the Open window, but each .fam's .rul file must also be present for the program to work.

**E-MAIL
FROM
DR. FUZZY**

If you have files in a directory other than the one containing the Knowledge Builder or on another disk drive, don't try to access them from within the Knowledge Builder. Instead, copy them from the Knowledge Builder directory, from the DOS prompt or in Windows.

Be sure to copy both the .rul and .fam file for each project.

The program allows manual control of information entry or a semi-automated "follow along" process.

As you saw in Chapter 3, before you start creating the knowledge base, you need the complete expert body of knowledge. For your introduction to the Fuzzy Knowledge Builder™, you'll use the scenario of creating an action graphic of a lunar lander with two Cartesian (*xyz* or length–width–depth) axes of motion. The lander can move up and down perpendicular to the surface. It also moves back and forth parallel to the moon's surface. This requires two separate knowledge bases that use the same input and output dimensions, but have different set values.

When you open the Fuzzy Knowledge Builder™, you'll face a blank screen with two menus—File and Help (Figure 4.1).

Figure 4.1: Opening screen.

**E-MAIL
FROM
DR. FUZZY**

The Fuzzy Knowledge Builder™ has a comprehensive hypertext Help system. You can click on the Help menu at any time, and get a complete listing of Help information. In addition, wherever you are in the program, you can get context-specific help.

After you click on File, your main choices are New and Open. Since this is a new project, click on New, which presents the full menu and, just below it, the icon toolbar (Figure 4.2). The icons are defined in Table 4.1.

**E-MAIL
FROM
DR. FUZZY**

At any time after you've named your project, you can save it by clicking on the Save icon, then closing the program in the usual Windows manner. To work on your file again, open the Fuzzy Knowledge Builder™, click on File and Open, then double-click on your project's filename. All filenames in the Open window have the extension .fam.

Dr. Fuzzy believes in hands-on learning, so in the rest of this chapter you'll use the Fuzzy Knowledge Builder™ to create knowledge bases for two "real-world" projects, a lunar lander and a personnel detection system. Then you'll have the opportunity to test them in an inference engine.

Figure 4.2: Full menu after a project has been defined.

TABLE 4.1: Icon Definitions.

		Definition	
		Click	*Double Click*
		Save file	Open file
		Display rule matrix	Control appearance
		Display rule box	None
		Select axis	Hidden select
		Copy edit	Paste edit
		Fuzzy set styles	Control appearance
		Open Action Tester	Recalculate view
		Build 3-D Viewer	Control appearance
		Build Gradient Viewer	Control appearance
		Build Profile Viewer	Control appearance
		Build formatted file	View formatted file
		Open all Edit windows	Close all windows
		Tile windows horizontal	Tile vertical

LUNAR LANDER

A lunar lander is a spacecraft designed to descend from a space ship to the moon's surface (and to return to the ship) and to fly across it. This means that the ship has two distinct movement axes, Vertical for descent and Horizontal for flying. This lander was meant as a computer simulation, not for an actual propulsion system. But it incorporates the physical features of an actual vehicle and also the moon's atmosphere and gravitational characteristics. For instance, the moon's gravity is a force on the lander, and the lander's propulsion system produces a counterforce against the lunar gravity.

The Vertical knowledge base will the first order of business. (Definitions for the Horizontal axis will be an adaptation of those for the Vertical.)

E-MAIL   This lunar lander graphics scenario was created by Thomas
FROM   Baker, formerly of TRW, and has been slightly modified for
DR. FUZZY   this book.

Lunar Lander's Vertical Axis

The lunar lander has two input dimensions and one output dimension for the Vertical axis and the same number for the Horizontal. The first step in knowledge base creation is naming and defining the dimensions. The input dimensions are Distance, measured in meters, and Velocity, measured in meters per second. The output dimension is Thrust, a force that's measured in meters per second.

Defining the System

Click on File and New from the opening menu. You'll be presented with a dialog box named New Fuzzy Node (Figure 4.3), which asks you to name the project. Type in the title **lunar_v**.

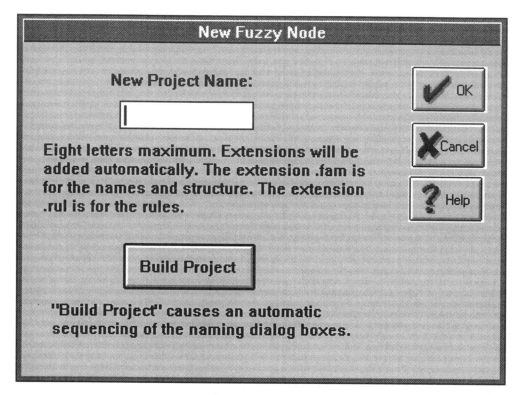

Figure 4.3: New Fuzzy Node dialog box.

You now have two options—open the definition screens one at a time or allow the Fuzzy Knowledge Builder™. We'll take the automatic route here, so click on the Build Project button.

First, name and describe the Input dimensions. The screen now displays the Input Dimension dialog box (Figure 4.4). Lunar-v requires two input dimensions, the maximum available with this version of Fuzzy Knowledge Builder™. Click on the radio button control on the left edge, which defines the number of input dimensions.

The next step is to fill in the basics about each dimension. Enter the information in Table 4.2, tabbing from field to field. If you need to return to a field, point and click with the mouse.

When you've finished, store these values and definitions by clicking on the OK button.

E-MAIL FROM DR. FUZZY

To open the screens individually instead, press Enter after you enter **lunar_v**. This returns you to the main screen. Now click on the Naming menu and select the individual items one at a time, starting at the top.

TABLE 4.2: Input Dimensions Data

Name	Description	Units Name	Min	Max
dis	Distance	meters	-300	+300
vel	Velocity	m/sec	-20	+20

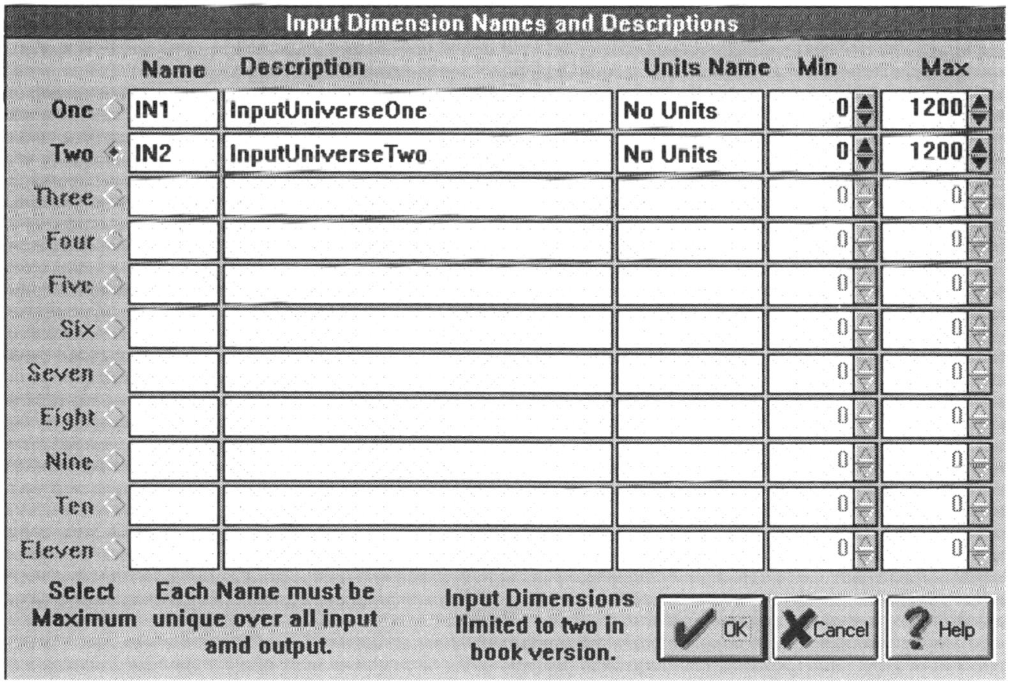

Input Dimension Names and Descriptions					
	Name	**Description**	**Units Name**	**Min**	**Max**
One	IN1	InputUniverseOne	No Units	0	1200
Two	IN2	InputUniverseTwo	No Units	0	1200
Three				0	0
Four				0	0
Five				0	0
Six				0	0
Seven				0	0
Eight				0	0
Nine				0	0
Ten				0	0
Eleven				0	0

Select Maximum   Each Name must be unique over all input amd output.   Input Dimensions limited to two in book version.   ✔ OK   ✗ Cancel   ? Help

Figure 4.4: Input dimension Names and Descriptions dialog box.

E-MAIL   For graphics purposes,
FROM
DR. FUZZY   1 meter = 1 pixel

Next, you must define the fuzzy membership functions (fuzzy sets) for each
dimension. The screen displayed in Figure 4.5 is the Input Fuzzy Sets Names
and Descriptions dialog box.

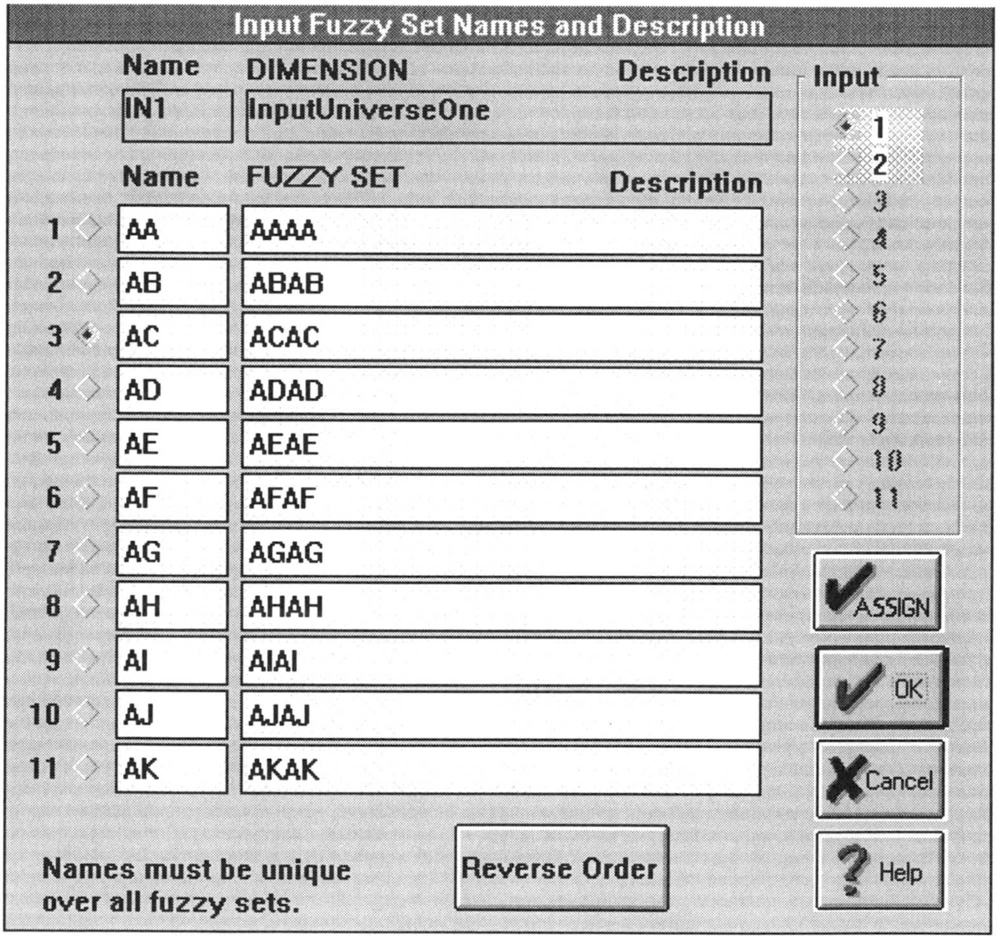

Figure 4.5: Input Fuzzy Sets Names and Descriptions dialog box.

TABLE 4.3: Lunar Lander Fuzzy Set Names and Descriptions

Name	Description
LN	Large negative
N	Negative
SN	Small negative
Z	zero
SP	Small positive
P	Positive
LP	Large positive

We want to define the Input dimensions. At the top are the name and description of the first input dimension, Distance. At the right, it's designated Input 1.

The next task is to define the number of fuzzy sets in this dimension, which is seven. Click on the button next to 7 in the left-hand column. Now enter the set names and descriptions as listed in Table 4.3.

E-MAIL FROM DR. FUZZY   All input and output dimensions in the lunar lander scenario have the same seven fuzzy set names, as given in Table 4.3. However, each dimension's sets have different numerical values.

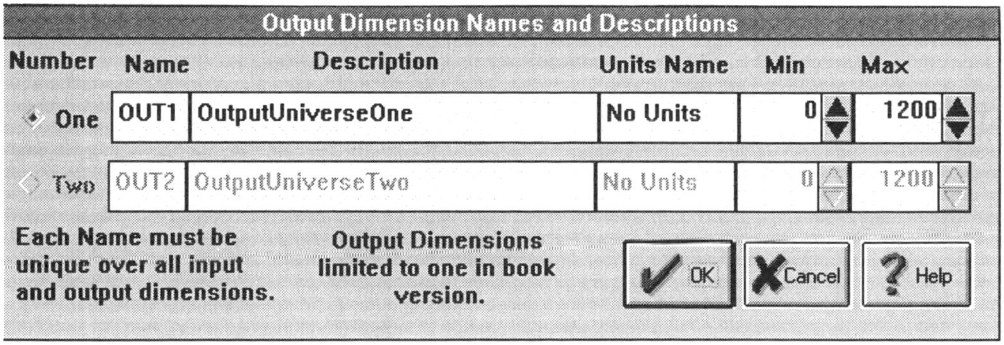

Output Dimension Names and Descriptions					
Number	**Name**	**Description**	**Units Name**	**Min**	**Max**
◆ One	OUT1	OutputUniverseOne	No Units	0 ▲▼	1200 ▲▼
◇ Two	OUT2	OutputUniverseTwo	No Units	0 △▽	1200 △▽

Each Name must be unique over all input and output dimensions.   **Output Dimensions limited to one in book version.**   ✔ OK   ✗ Cancel   ? Help

Figure 4.6: Output dimension Names and Descriptions dialog box.

When you've finished entering the data for this dimension, click on the Assign button, to register what has just been entered. The number 2 input dimension box will be displayed. Fill it in with the same list of set names and descriptions. When you've finished, click on the Assign and OK buttons.

E-MAIL FROM DR. FUZZY   Assigning values *does not* save your data to the disk. *You can save the file at any time by clicking on the Save icon.*

The Output dimension and fuzzy set must also be named and described. The next screen displays the Output dimension dialog box (Figure 4.6). Follow the same procedure as with the Input dimension, entering the information in Table 4.4. When you are finished, click on OK. Fill in the next screen, Output Fuzzy Set Names and Descriptions, just as you did for the Input sets, using the data in Table 4.3. When you've finished, click on the OK and Assign buttons. The screen will then display a dialog box notifying you that RESIZING RULE MATRIX WILL INITIALIZE RULES! Click OK on this box and then on Assign.

TABLE 4.4: Output Dimension Data

Name	Description	Units Name	Min	Max
thr	Thrust	m/sec	-20	+20

E-MAIL FROM DR. FUZZY   CAUTION. Setting up the Input fuzzy sets defines the size of the rules matrix. If you decide to change the number of fuzzy sets after this point, you'll lose any work you've done on the rules matrix and have to do it all over again.
        Changing the shape or values of fuzzy sets doesn't destroy the rules matrix.

Congratulations! You've named and defined your dimensions and fuzzy sets. Now's a good time to save what you've done so far.

To take a look at the sets and rules matrix, click on the Display All Windows icon, which displays them in vertical format (Figure 4.7). To display them in horizontal format (Figure 4.8), click on the Tiles icon.

Defining the Rules

The next task involves the details of knowledge base design—defining the rules and associated fuzzy sets. The Fuzzy Knowledge Builder™ produces an evenly spaced series of fuzzy sets for each dimension. It generates the rules to match the fuzzy sets. You can adjust both the fuzzy sets and the matrix to meet the needs of your project.

Because each Input dimension has seven fuzzy sets, the rules matrix consists of 49 rules. Table 4.5 contains the scenario-defined list of rules.

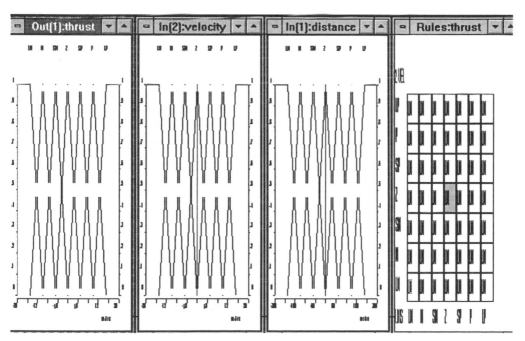

Figure 4.7: All windows open, vertical format.

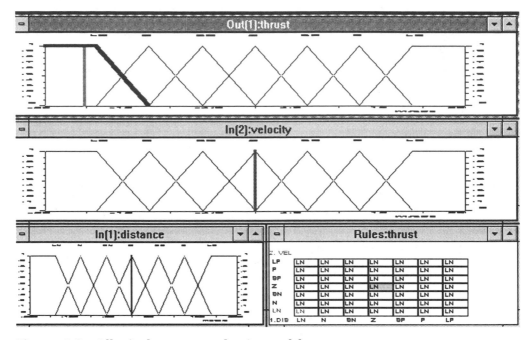

Figure 4.8: All windows open, horizontal format.

E-MAIL FROM DR. FUZZY

You can customize the appearance, the color, or both of virtually every aspect of the matrix and set displays. Double-click on the Matrix icon to display the matrix dialog box (Figure 4.9). Clock on Set Shape Style (Figure 4.10a) and Shaper Display (Figure 4.10b) in the sets menu for the set dialog boxes.

Generating the Rules Matrix and Rule Box

To begin the customization of the knowledge base rules matrix and rule box, generate each of them. Editing the rules involves the interaction of the Rules Matrix and the Rule Box.

TABLE 4.5: Scenario-Defined Rules for Lunar_v.

As Distance is Large Negative *and* Velocity is Large Negative, *then* Thrust is Zero

As Distance is Large Negative *and* Velocity is Negative, *then* Thrust is Zero

As Distance is Large Negative *and* Velocity is Small Negative, *then* Thrust is Zero

As Distance is Large Negative *and* Velocity is Zero, *then* Thrust is Zero

As Distance is Large Negative *and* Velocity is Small Positive, *then* Thrust is Zero

As Distance is Large Negative *and* Velocity is Positive, *then* Thrust is Zero

As Distance is Large Negative *and* Velocity is Large Positive, *then* Thrust is Zero

As Distance is Negative *and* Velocity is Large Negative, *then* Thrust is Zero

As Distance is Negative *and* Velocity is Negative, *then* Thrust is Zero

As Distance is Negative *and* Velocity is Small Negative, *then* Thrust is Zero

As Distance is Negative *and* Velocity is Zero, *then* Thrust is Zero

As Distance is Negative *and* Velocity is Small Positive, *then* Thrust is Zero

As Distance is Negative *and* Velocity is Positive, *then* Thrust is Zero

As Distance is Negative *and* Velocity is Large Positive, *then* Thrust is Zero

As Distance is Small Negative *and* Velocity is Large Negative, *then* Thrust is Zero

As Distance is Small Negative *and* Velocity is Negative, *then* Thrust is Zero

As Distance is Small Negative *and* Velocity is Small Negative, *then* Thrust is Zero

As Distance is Small Negative *and* Velocity is Zero, *then* Thrust is Zero

As Distance is Small Negative *and* Velocity is Small Positive, *then* Thrust is Zero

As Distance is Small Negative *and* Velocity is Positive, *then* Thrust is Zero

As Distance is Small Negative *and* Velocity is Large Positive, *then* Thrust is Zero

As Distance is Zero *and* Velocity is Large Negative, *then* Thrust is Zero

As Distance is Zero *and* Velocity is Negative, *then* Thrust is Zero

As Distance is Zero *and* Velocity is Small Negative, *then* Thrust is Zero

As Distance is Zero *and* Velocity is Zero, *then* Thrust is Small Positive

As Distance is Zero *and* Velocity is Small Positive, *then* Thrust is Small Positive

As Distance is Zero *and* Velocity is Positive, *then* Thrust is Zero

As Distance is Zero *and* Velocity is Large Positive, *then* Thrust is Zero   *(continued)*

TABLE 4.5: Scenario-Defined Rules for Lunar_v *(continued).*

As Distance is Small Positive *and* Velocity is Large Negative, *then* Thrust is Zero

As Distance is Small Positive *and* Velocity is Negative, *then* Thrust is Zero

As Distance is Small Positive *and* Velocity is Small Negative, *then* Thrust is Zero

As Distance is Small Positive *and* Velocity is Zero, *then* Thrust is Small Positive

As Distance is Small Positive *and* Velocity is Small Positive, *then* Thrust is Small Positive

As Distance is Small Positive *and* Velocity is Positive, *then* Thrust is Small Positive

As Distance is Small Positive *and* Velocity is Large Positive, *then* Thrust is Zero

As Distance is Positive *and* Velocity is Large Negative, *then* Thrust is Zero

As Distance is Positive *and* Velocity is Negative, *then* Thrust is Zero

As Distance is Positive *and* Velocity is Small Negative, *then* Thrust is Zero

As Distance is Positive *and* Velocity is Zero, *then* Thrust is Positive

As Distance is Positive *and* Velocity is Small Positive, *then* Thrust is Positive

As Distance is Positive *and* Velocity is Positive, *then* Thrust is Positive

As Distance is Positive *and* Velocity is Large Positive, *then* Thrust is Small Positive

As Distance is Large Positive *and* Velocity is Large Negative, *then* Thrust is Zero

As Distance is Large Positive *and* Velocity is Negative, *then* Thrust is Zero

As Distance is Large Positive *and* Velocity is Small Negative, *then* Thrust is Zero

As Distance is Large Positive *and* Velocity is Zero, *then* Thrust is Large Positive

As Distance is Large Positive *and* Velocity is Small Positive, *then* Thrust is Positive

As Distance is Large Positive *and* Velocity is Positive, *then* Thrust is Positive

As Distance is Large Positive *and* Velocity is Large Positive, *then* Thrust is Positive

Click on the Rules Matrix item in the toolbar to generate the rules matrix, then click on the Rule Box item of the Display submenu of the Rules menu bar to generate the Rule Box. After you've generated them, you can display them with their toolbar icons.

Display the Rules Matrix (Figure 4.11) by clicking on its icon. Place the cursor on any rule (cell). Then you can display its rule in full by clicking

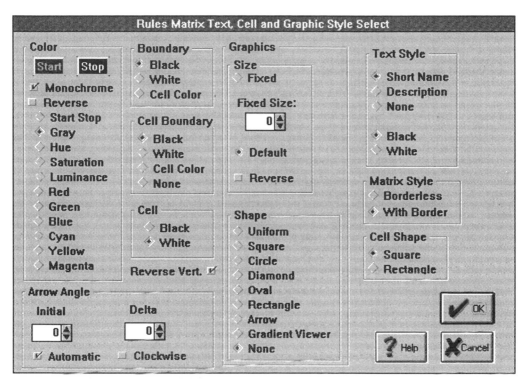

Figure 4.9: Rules Matrix Display Style box.

on the Display rule box icon. The Rule Box is movable—just point and drag it with the mouse.

Notice that putting the mouse cursor over a cell highlighted it, turning the cell characters red, the associated fuzzy set name blue, and the associated vertical fuzzy set name green. Pressing the Shift key while moving the mouse will stop the highlighting changes.

To move the cursor from the matrix to the rule box, select a cell and press the Shift key. The cursor will jump to the rule box.

(a)

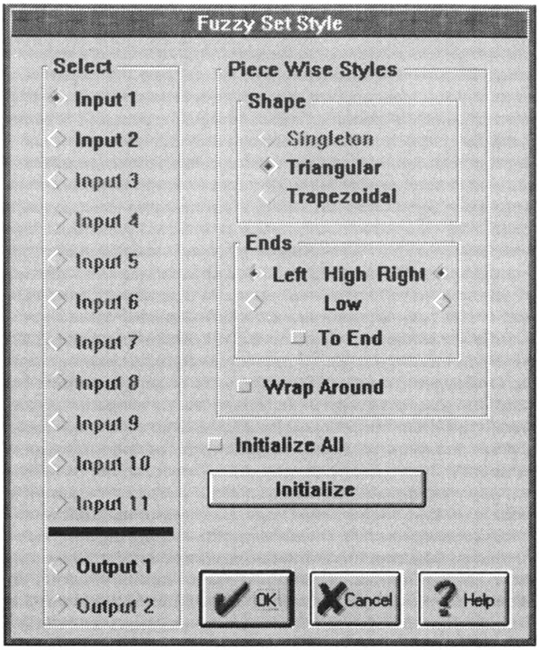

(b)

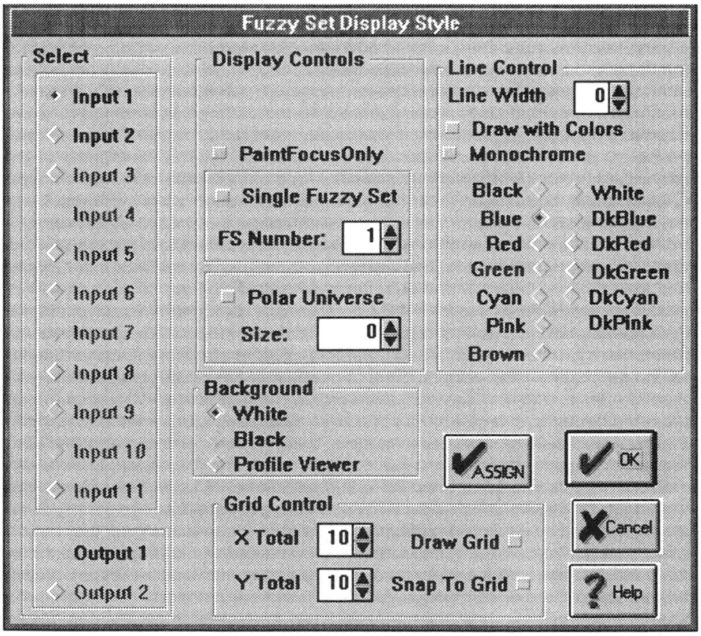

Figure 4.10: Fuzzy Set Display dialog box.

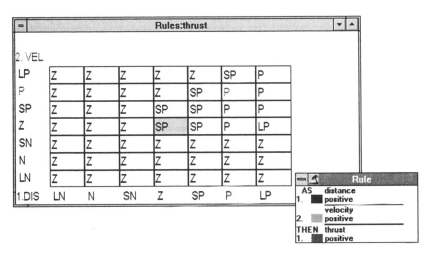

Figure 4.11: Rules Matrix and Rule Box.

	Where should you start your editing? Think of it in terms of

E-MAIL   Where should you start your editing? Think of it in terms of
FROM   coarse or big-change editing and fine-tune editing. To make
DR. FUZZY   big modifications in a few steps, adjust the rules. For fine-
tuning, adjust the numerical values of the membership functions (fuzzy sets).

Using the Rules Matrix and Rule Box

The end goal of any rule change is to provide an improved output value from the interaction of input values with the rules matrix. To change a rule in the Rules Matrix, place the cursor over the selected cell and click with the *left* mouse button. This will change the rule to the next one in the list—that is, to the next fuzzy Output set to the right. For instance, if the cells is LN, it will change to Z.

Clicking on a cell with the *right* mouse button will change it to the previous one in the list—to the next Output fuzzy set to the left. For example, if the cell is SP, it will change to P.

You can also change the rule directly, in the Rule Box. Move the cursor over the *then* portion of the desired rule. Click on it with the left mouse button to move to the next Output fuzzy set; click with the right mouse button to move to the previous one.

When you've adjusted the rules to the requirements of the scenario, save your work. Next, you should fine-tune your project by editing the fuzzy membership functions (sets) directly.

Editing the Fuzzy Sets

The next step in the design of the fuzzy action surface is to define the fuzzy sets or membership functions for the application. Table 4.6 contains the scenario's values for the Vertical axis's fuzzy sets.

Display the fuzzy sets by clicking on the Display All Windows icon. For horizontal display, click on the Tile icon. To enlarge one of the sets, click on the UP button in its upper right-hand corner. Or choose the Input sets (Input Set Shapers) or Output set (Output Set Shapers) from the Sets menu.

Move the cursor close to the segment end point of interest. Then press and hold the left mouse button. The cursor will move to the end point if close enough. Continue to hold the button while moving the cursor. The end point will follow the cursor. The end point will stay at its last location when the button is released. Movement is contrained to the high or low lines.

Sometimes an end point is hard to select because it occupies the same place as another end point. To make selection easier, hold down the Shift key while pressing the left mouse button.

If you want to move an entire fuzzy set on the universe of discourse, press Ctrl when selecting one of its end points.

Be sure to save your work frequently.

Examining the Knowledge Base

The Fuzzy Knowledge Builder™ provides several methods for statically testing and viewing the action surface you have designed—an Action Tester, a 3-D Viewer, a Gradient Viewer, and a Profile Viewer.

The static testing tool is called the *Action Tester*. It accepts crisp values over the input dimensions and processes them through the action surface.

TABLE 4.6: Scenario-Supplied Values for the Vertical Fuzzy Sets

	Left	Center	Right
Velocity:			
LN	-20	-10	-5
N	-7	-4	-1
SN	-2	-1	0
Z	-.5	0	.5
SP	0	1	2
P	1	4	7
LP	5	10	20
Distance:			
LN	-300	-200	-100
N	-140	-80	-20
SN	-40	-20	0
Z	-10	0	10
SP	0	20	40
P	20	80	140
LP	100	200	300
Thrust:			
LN	-20	0	0
N	0	0	0
SN	0	0	0
Z	0	0	.12
SP	0	.1	.24
P	.1	.6	1.08
LP	.6	1.2	-20

You can use this tool at any time to test a knowledge base that you're building.

Click on the Action Tester or in the Tools submenu or in the toolbar. The Action Tester dialog will appear with the dimensions and scaling as defined for the Open design (Figure 4.12).

Each input dimension is represented by a scroll bar. The crisp input value may be set by positioning the scroll thumb with the mouse cursor (press left button over thumb, move cursor dragging thumb, release button to leave thumb in new position).

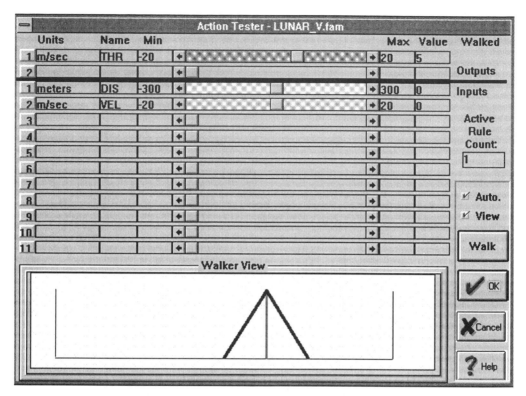

Figure 4.12: Action Tester.

The scaled crisp value of where the thumb is along the input dimension appears immediately right of the scroll bar. Normal keyboard commands will position the thumb, as well.

Enter all values for all active dimensions. If the Dynamic check box is not checked, click on the Map button. In a short time the Output scrollbar thumbs will be positioned as the Knowledge Builder passes your Input values through the action surface.

The top scrollbar provides the crisp Output.

The 3-D Viewer provides a 3-D look at the Output plotted against Input dimensions (see Figure 4.13).

Figure 4.14 provides the perspective of the Gradient Viewer showing Output plotted against Input dimensions.

Figure 4.15 provides the perspective of the Profile Viewer showing Output plotted against Input dimensions.

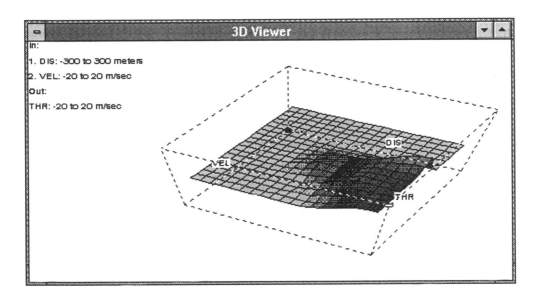

Figure 4.13: A 3-D view of the Input–Output plot.

Now you've been through the entire process of creating a knowledge base for the lunar lander's Vertical axis. You can create a knowledge base for the Horizontal axis by adapting the work that you've just done.

Lunar Lander's Horizontal Axis

Begin creating the Horizontal axis's knowledge base by saving the lunar_v.fam file as *lunar_h.fam*. Click on File and SaveAs.

Next, adjust the rules according to the rules listed in Table 4.7. You can do this on the Rules Matrix or in the Rule Box. When you've finished, save your work.

Now look at the fuzzy sets. You can adjust them so they have the values in Table 4.8.

Finally, test and review the knowledge base with the Tools described in the last section.

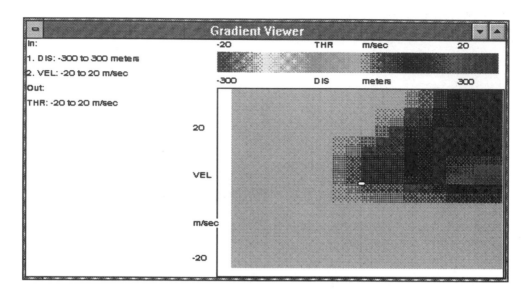

Figure 4.14: A gradient view of the Input–Output plot.

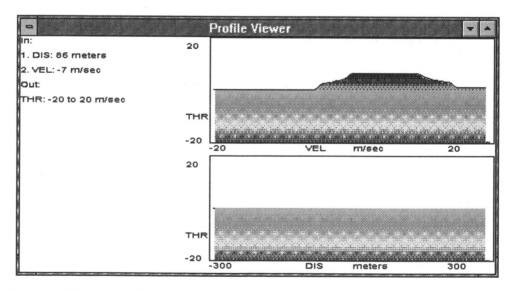

Figure 4.15: A profile view of the Input–Output plot.

TABLE 4.7: Scenario-Defined Rules for Lunar_h.

As Distance is Large Negative *and* Velocity is Large Negative, *then* Thrust is Zero

As Distance is Large Negative *and* Velocity is Negative, *then* Thrust is Small Negative

As Distance is Large Negative *and* Velocity is Small Negative, *then* Thrust is Negative

As Distance is Large Negative *and* Velocity is Zero, *then* Thrust is Negative

As Distance is Large Negative *and* Velocity is Small Positive, *then* Thrust is Negative

As Distance is Large Negative *and* Velocity is Positive, *then* Thrust is Large Negative

As Distance is Large Negative *and* Velocity is Large Positive, *then* Thrust is Large Negative

As Distance is Negative *and* Velocity is Large Negative, *then* Thrust is Small Positive

As Distance is Negative *and* Velocity is Negative, *then* Thrust is Zero

As Distance is Negative *and* Velocity is Small Negative, *then* Thrust is Small Negative

As Distance is Negative *and* Velocity is Zero, *then* Thrust is Small Negative

As Distance is Negative *and* Velocity is Small Positive, *then* Thrust is Negative

As Distance is Negative *and* Velocity is Positive, *then* Thrust is Negative

As Distance is Negative *and* Velocity is Large Positive, *then* Thrust is Large Negative

As Distance is Small Negative *and* Velocity is Large Negative, *then* Thrust is Small Positive

As Distance is Small Negative *and* Velocity is Negative, *then* Thrust is Small Positive

As Distance is Small Negative *and* Velocity is Small Negative, *then* Thrust is Small Positive

As Distance is Small Negative *and* Velocity is Zero, *then* Thrust is Small Negative

As Distance is Small Negative *and* Velocity is Small Positive, *then* Thrust is Small Negative

As Distance is Small Negative *and* Velocity is Positive, *then* Thrust is Negative

As Distance is Small Negative *and* Velocity is Large Positive, *then* Thrust is Negative

As Distance is Zero *and* Velocity is Large Negative, *then* Thrust is Positive

As Distance is Zero *and* Velocity is Negative, *then* Thrust is Small Positive

As Distance is Zero *and* Velocity is Small Negative, *then* Thrust is Small Positive

As Distance is Zero *and* Velocity is Zero, *then* Thrust is Zero

As Distance is Zero *and* Velocity is Small Positive, *then* Thrust is Small Negative

As Distance is Zero *and* Velocity is Positive, *then* Thrust is Small Negative

As Distance is Zero *and* Velocity is Large Positive, *then* Thrust is Negative   *(continued)*

TABLE 4.7: Scenario-Defined Rules for Lunar_h *(continued)*.

As Distance is Small Positive *and* Velocity is Large Negative, *then* Thrust is Positive

As Distance is Small Positive *and* Velocity is Negative, *then* Thrust is Positive

As Distance is Small Positive *and* Velocity is Small Negative, *then* Thrust is Small Positive

As Distance is Small Positive *and* Velocity is Zero, *then* Thrust is Small Positive

As Distance is Small Positive *and* Velocity is Small Positive, *then* Thrust is Zero

As Distance is Small Positive *and* Velocity is Positive, *then* Thrust is Small Negative

As Distance is Small Positive *and* Velocity is Large Positive, *then* Thrust is Small Negative

As Distance is Positive *and* Velocity is Large Negative, *then* Thrust is Large Positive

As Distance is Positive *and* Velocity is Negative, *then* Thrust is Positive

As Distance is Positive *and* Velocity is Small Negative, *then* Thrust is Positive

As Distance is Positive *and* Velocity is Zero, *then* Thrust is Small Positive

As Distance is Positive *and* Velocity is Small Positive, *then* Thrust is Small Positive

As Distance is Positive *and* Velocity is Positive, *then* Thrust is Zero

As Distance is Positive *and* Velocity is Large Positive, *then* Thrust is Small Negative

As Distance is Large Positive *and* Velocity is Large Negative, *then* Thrust is Large Positive

As Distance is Large Positive *and* Velocity is Negative, *then* Thrust is Large Positive

As Distance is Large Positive *and* Velocity is Small Negative, *then* Thrust is Positive

As Distance is Large Positive *and* Velocity is Zero, *then* Thrust is Positive

As Distance is Large Positive *and* Velocity is Small Positive, *then* Thrust is Positive

As Distance is Large Positive *and* Velocity is Positive, *then* Thrust is Small Positive

As Distance is Large Positive *and* Velocity is Large Positive, *then* Thrust is Zero

Printing Your Graphics Displays

The Fuzzy Knowledge Builder™ lets you print any graphics window that's displayed on the screen. The printer will reproduce what's on the screen, so make it the desired size before beginning the printing process.

TABLE 4.8: Scenario-Supplied Values for the Horizontal Fuzzy Sets

	Left	*Center*	*Right*
Velocity:			
LN	-20	-12	-6
N	-8.4	-4.8	-1.2
SN	-2.4	-1.2	0
Z	-.6	0	.6
SP	0	1.2	2.4
P	1.2	4.8	8.4
LP	6	12	20
Distance:			
LN	-300	-200	-100
N	-140	-80	-20
SN	-40	-20	0
Z	-10	0	10
SP	0	20	40
P	20	80	140
LP	100	200	300
Thrust:			
LN	-20	-8	-4
N	-8	-4.8	-.8
SN	-1.6	-.8	0
Z	-.4	0	.4
SP	0	.8	1.6
P	.8	4.8	8
LP	4	8	-20

Click on the File menu and then on Print. The dialog box in Figure 4.16 will appear on the screen. Select the set or matrix by clicking on its button and then on OK.

Another way to print your graphic is to place it in the Windows clipboard by pressing Print Screen (for the entire screen) or Alt-Print Screen (for the just active window). You can then paste the graphic into Write, Word, or other document editor.

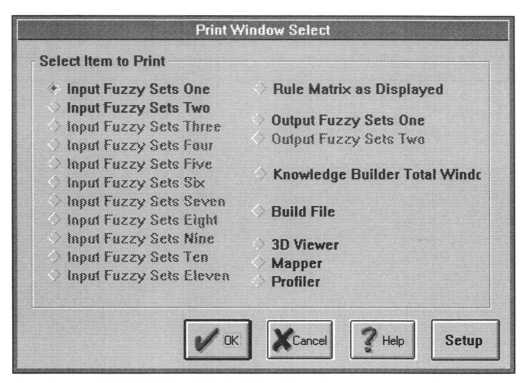

Figure 4.16: Print Window Select dialog box.

PERSONNEL DETECTION SYSTEM

Is someone in the building or not? This is a question that automated personnel detection systems are designed to answer. In office buildings, warehouses, and military installations, human security guards are augmented by infrared, sound, photoelectric, and other kinds of sensors. In the opinion of many experts, existing sensor systems aren't particularly accurate. Dr. Paul Sayka, of Los Alamos National Laboratory, has proposed a superior system using fuzzy logic. Because the system has two input dimensions and one output, it's an excellent example for knowledge base construction by the Fuzzy Knowledge Builder™.

Naming and Defining the Dimensions and Sets

Begin by selecting New from the File menu. Name the project Sensors, then begin defining the input and output sets through the Naming menu, using the data in Table 4.9. One input dimension is an optical sensor whose fuzzy sets are named *Slow, Medium,* and *Fast.* It is measured in miles per hour from 0 to 9.

    The other input dimension is an infrared sensor, with fuzzy sets named *Very Low, Low, Medium, High,* and *Very High.* Its units are temperatures from 94° to 108° Fahrenheit.

    The output dimension is an audible signal with adjustable-length sound pulses. Its fuzzy sets are named *Very Low, Low, Medium,* and *Very High,* measured in milliseconds from 500 to 2000.

    Tables 4.10 and 4.11 contain the data to be entered in the name-and-description tables.

    Make any necessary adjustments to the membership functions, which can be displayed by clicking on the Open All Windows icon or selecting Set Shapers from the Sets menu. To display one set at a time for editing, double-click on the Fuzzy Set Display icon, click on the Single Fuzzy Set button, and choose the set number (1 is the leftmost set). Then click on Assign and OK.

    The rule matrix should look like the one in in Table 4.12.

TABLE 4.9: Sensors Dimensions Data.*

Name	*Description*	*Units Name*	*Min*	*Max*
Input:				
opt	Optical sensor	MPH	0	9
ir	Infrared	Degrees F	900	1050
Output:				
sgn	Audible signal	Milliseconds	500	2000

* Data from Paul Sayka, "A Fuzzy Logic Rule-Based System for Personnel Detection," in *Fuzzy Logic and Controls: Software and Hardware Applications,* Jamshidi/Vadiee/Ross, eds., © 1993, pp. 227–229. Reprinted by permission of Prentice Hall, Englewood Cliffs, New Jersey.

TABLE 4.10: Input Fuzzy Sets' Data*

		Values		
Name	Description	Left	Center	Right
Optical:				
slo	Slow	0	1.5	3
med	Medium	2	4	6
fast	Fast	5	7	9
Infrared:				
vlo	Very Low	94	96	98
low	Low	95	97	99
med	Medium	98	100	102
high	High	101	103	105
vhi	Very High	104	106	108

TABLE 4.11: Audible Signal (Output) Fuzzy Set Names and Descriptions.

Name	Description
vlo	Very low
low	Low
med	Medium
high	High

TABLE 4.12: Personnel Detection Rules Matrix.

Infrared	VLO	LOW	MED	HIGH	VHI
O SLO	VLO	LOW	VLO	LOW	VLO
p					
t					
i MED	LOW	MED	LOW	MED	LOW
c					
a					
l FAST	MED	HIGH	MED	LOW	VLO

* Data for Tables 4.10–4.12 from Paul Sayka, "A Fuzzy Logic Rule-Based System for Personnel Detection," in *Fuzzy Logic and Controls: Software and Hardware Applications*, Jamshidi/Vadiee/Ross, eds., © 1993, pp. 227–229. Reprinted by permission of Prentice Hall, Englewood Cliffs, New Jersey.

Improving the Matrix's Operation

Put yourself in the expert's place. Examine the sets with the viewers and try them out in the Tester. Are the output signals what you want in *your* personnel detection system? How do you tell a "good" action surface from a "not so good" one?

The best surface is one that flows the most smoothly from one area to another—it has the smoothest gradient. That's what all the viewers can tell you.

E-MAIL
FROM
DR. FUZZY

Want to see what a really "unsmooth" matrix gradient looks like? Open the 3-D Viewer and select Random. You'll think you've been transported to the Alps.

The slogan here is "random in, random out." You want your matrix to be as unrandom as possible.

You may want to adjust the rules matrix directly by clicking on individual matrix cells. You can also use the Knowledge Builder's automatic correction tools, collected as the Knowledge Wizard (Figure 4.17). They are available from the Rules menu.

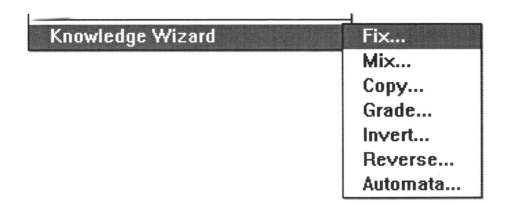

Figure 4.17: Knowledge Wizard menu.

Automata

The major Wizard tool is called *Automata,* a team of small programs called *cellular automata* (CAs) that make the matrix cell boundaries fuzzier and smoother by either inserting (interpolating) rules between existing ones or smoothing (annealing) the overall surface. This produces an action surface that's closer to the ideal one that you want for your system's outputs.

You have the option of designating any existing matrix rule as Fixed (nonadjustable). Just hold down the Control key while clicking on the rule.

The Manifold Automata selection box (Figure 4.18) lets you choose the mathematical formula (algorithm) that will control the CA actions. The Limit Cycles button lets you choose the number of cycles that the CAs will undergo while improving the rules.

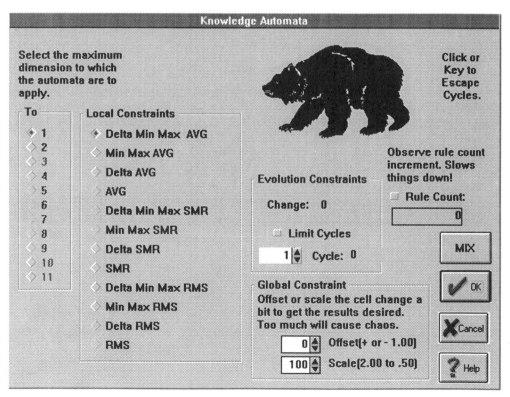

Figure 4.18: Automata selection box.

The Global Constraint area lets you enter a "fudge factor" in the form of either a globally applied linear offset or a gain (scaling) of the effect that adjacent state changes have on a particular rule.

If you click on the Mix button, the output values are automatically randomized over all the rules or over those rules that haven't been designated as Fixed.

The Fix feature is also available directly from the Wizard menu. It also lets you designate a rule as "don't care," meaning it doesn't matter whether it is Fixed or Not Fixed.

E-MAIL FROM DR. FUZZY

Each cellular automaton, or CA, is a "machine" that changes its state (such as ON or OFF) according to a set of rules that affect it and its neighbors.

For instance, a CA rule may state that if three of a CA's neighbors are ON, it will be ON. Another rule may state that if four of its neighbors are ON, the CA will be OFF. The rules apply equally to each CA in the matrix. This means each CA's state is determined by its neighbors' states and that its own state partially determines the state of each of its neighbors.

The Fuzzy Knowledge Builder™'s CAs have as many states as there are output sets. You can choose which of a series of algorithms will be used to perform the state changes in the Automata selection box (Figure 4.18).

Click for Help, and the equations will be displayed.

Other Wizard Tools

Other tools are also available through the Rules menu's Manifold Wizard item. Grade provides automatic generation of gradients (hills) in the rules matrix. The dialog box lets you choose whether the slope is positive or negative. The To area lets you choose the slope of the gradient. If you want a flat ("city block") measurement of distances, click on the Flat style. For a curved gradiant, select either RMS (root mean square) or SMR (square mean root).

FORMATTING THE KNOWLEDGE BASE FOR AN INFERENCE ENGINE

Now that you've constructed, tested, and tweaked your knowledge base, you can use the Build menu to format it for an inference engine (Figure 4.19). The book version of the Fuzzy Knowledge Builder™ provides four formats: analog devices, C language, Motorola 68HC05 processor, and Scripts (for QBASIC).

For instance, you can build the Sensors knowledge base into a Scripts file (Figure 4.20) and a C language file (Figure 4.21).

To create a Scripts file, select Scripts from the Build menu, displaying the Data File Style box (Figure 4.22).

Click on OK to select the default values, and the Fuzzy Knowledge Builder™ will produce a Scripts file with the name of your project plus the extension *.FDT*. You can display the file immediately by selecting View Built File from the Build menu, or read it through Windows Write, the Notepad, or another text editor or word processor that you select in the Naming menu's

Figure 4.19: Build menu item of menu bar.

```
─ 🎨              Notepad - SENSORS.FDT
 File  Edit  Search  Help
SENSORS.fdt
Date: 1994-4-30 Time: 14:47:54
NUM_INPUTS
2
NUM_OUTPUTS
1
CREDIBILITY
0
INPUT1
3
OPT
SLOW,MED,FAST
INPUT2
5
IR
VLOW,LOW,MED,HIGH,VHI
OUTPUT1
4
```

Figure 4.20: Portion of a Scripts formatting of the Sensors knowledge base.

```
─ 🎨              Notepad - SENSORS.FIC
 File  Edit  Search  Help
/*
* SENSORS.fic
*/
/*
**    Date: 1994-4-30 Time: 14:51:6    */
/*
**      C Language include file.
**      Generated by Fuzzy Knowledge Builder.
**      Fuzzy Systems Engineering.
**      Copyright 1991, 1993, 1994 .
**      This Fuzzy Knowledge Builder copy is
**      registered for the single computer use by:
**      Fuzzy Logic: A Practical
**      Approach, McNeill and Thro
**      (C) 1994 Academic Press
```

Figure 4.21: Portion of a C language formatting of the Sensors knowledge base.

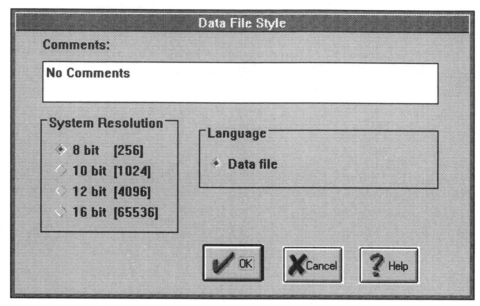

Figure 4.22. Data File Style box.

Options dialog box. You can also copy it in Windows and paste it into any other document. Or double-click on the View Knowledge Base icon to display it.

Depending on your project's requirements, you may want to select another system or variable resolution. For example, the dialog box for C language files allows choices of variable resolution for inputs and outputs, Fuzzy One, and variable type.

For more on inference engines that run these knowledge base files, see Appendix D.

USING A KNOWLEDGE BASE IN AN INFERENCE ENGINE

Dr. Fuzzy has provided two extremely simple, but functional, DOS-based inference engines written in QBASIC named *fuzzy1.bas* and *fuzzy2.bas*. They both support triangular and trapezoidal membership functions. Fuzzy1 tests all rules. Fuzzy2 is slightly faster, because it tests only active rules.

Before running your own knowledge base through either of the inference engines, you may want to test drive fuzzy1 with a file named (appropriately) *test.fdt*. Return to a DOS command line and copy QBASIC into the same directory as the knowledge base files.

Then at the command prompt, type

qbasic /run fuzzy1.bas

and Enter. (Or use fuzzy2 instead.)

From the opening menu, press *R*, so that the test.fdt knowledge base is read into the inference engine. You'll see it whiz across the screen in the process.

Next, press *I*, and enter input values between 0 and 255 for Input One and Input Two, such as

Input One: 50

and Enter, then

Input Two: 120

and Enter. Then press *Y* to return to the opening menu.

Now perform the actual inference operation by typing *F*. The screen will display the output value, which is 75 for the input values you entered. Pressing *Y* returns to the opening menu, and you can press *Q* to exit the inference engine.

You can see the anatomy of the inference process in two ways. If you're a QBASIC programmer or a wannabee, you can turn to Appendix D and examine the printouts of the fuzzy1 and fuzzy2 code. You can also return to the Fuzzy Knowledge Builder™, load the file *test.fam*, and examine the inputs 50 and 120 in the Action Tester. You'll see that the output value agrees closely with that provided by fuzzy1 or fuzzy2. By opening the rules matrix and the rule box, you can see which rules are active and their text.

Now that you've driven the inference engine around the block, you can put it to work on another file, such as the Sensors file.

Fuzzy1 and fuzzy2 have the filename test.fdt written into the code, so you have two choices. You can change the name of your own file to test.fdt. Or you can go into the inference engine and change those lines. For instance, change test.fdt to *sensors.fdt*. Three places in either engine require editing.

The Appendix D printouts display those lines in bold type, to make the task easier.

E-MAIL
FROM   Intrigued by inference engines? Appendix D offers other in-
DR. FUZZY   ference engines and test files for you to play with.

This completes the coverage of fuzzy expert systems. But other Fuzzy tools await. The next chapter introduces the Fuzzy Decision Maker.

CHAPTER 5

DESIGNING A FUZZY DECISION

Decisions! We make them all the time. Should I buy a house or rent an apartment? What's the best deal—buying a new car, buying a used one, or leasing one? Should I take a better-paying job in another state or stay put? Should I marry the person I'm dating? Will we be happy together? Which college should I go to? Should I buy stocks or bonds? Should I buy a new computer or get my present one upgraded? What's the most effective strategy for selling our new product?

We make personal and business decisions all the time. What they have in common is the process we go through. The answer's hardly ever cut and dried or a clear Yes or No. When you really examine it, even the "simplest" decision can turn out to be quite complex. In other words, the situations are fuzzy, and so is the way we make our choices.

Unlike the fuzzy expert system, fuzzy decision making is just starting to be used in the real world. As people are beginning to learn, it's a useful tool. The Fuzzy Decision Maker™ is the first commercial fuzzy decision-making application.

E-MAIL FROM DR. FUZZY

The Fuzzy Decision Maker™ uses a decision method developed by Dr. Michael O'Hagan and Nadja K. O'Hagan (Fuzzy Logic, Inc.), based on prior work by Ronald K. Yager, Thomas L. Saaty, and other researchers. It has been used by the U.S. Department of the Army for the National Guard. (The theory is discussed in the 1993 paper by O'Hagan and O'Hagan listed in the Bibliography.)

The Fuzzy Decision Maker™'s implementation of this theory is described in the last section of this chapter.

In this chapter, you'll use the Fuzzy Decision Maker™ on three decision scenarios. One involves selecting a college. A second one is government oriented, choosing the best way to optimize a regional traffic system. Finally, you'll adapt the decision process to evaluate the pros and cons of a proposed merger—in this case, marriage.

THE DECISION PROCESS

Three kinds of information go into the decision process—Goals, Constraints, and Alternatives. *Goals* are what we want out of the process, such as an affordable home, a reliable way to get to work, a satisfying job, a long-lasting relationship, a memorable college education. *Constraints* are limiting factors, such as the amount of money we have to spend, the geographical area we prefer, or spouse's preferences. *Alternatives* are the available choices—private college or public, private car or mass transit, salary or commission.

The Fuzzy Knowledge Builder™ transforms continuous variables into other continuous variables. The Fuzzy Decision Maker™ transforms discrete concepts into other discrete concepts. For instance, it uses unconnected goals and constraints with unconnected alternatives to determine the most valuable alternative.

There are several recognized methods of handling Goals, Constraints, and Alternatives in a decision-making system. Such a system is designed to produce the decision (choose the Alternative) that best meets the Goals, within the bounds of the Constraints. The Fuzzy Decision Maker™'s method gives ranked Goals and ranked Constraints the same importance in the decision process. The decision is arrived at by the process of aggregation, in which the ranked Goals and Constraints are reduced to a single number.

INTRODUCING THE FUZZY DECISION MAKER™

The Fuzzy Decision Maker™ is designed to help a person make a decision that resembles a "do-it-myself" decision process. It even requires the person to recognize his or her unique preconceptions or unconscious biases and incorporate them in the decision process in an apparent and open way. As in noncomputerized real life, calculations are based on a whole series of little decisions about importances and satisfactions.

Each Goal, Constraint, and Alternative is weighed against the others somewhat in isolation. This includes their importance to you, their relative importances, and the degree to which each Alternative satisfies the Goals and Constraints. It breaks the decision scenario down into small parts that you can focus on and easily enter into the program. Then a special form of aggregation, based on the user's personal optimism–pessimism bias, is employed to rank the Alternatives.

E-MAIL FROM DR. FUZZY   The book version of the Fuzzy Decision Maker™ allows a maximum of four Goals, four Constraints, and four Alternatives. The commercial version allows many more of each category.

First, you must analyze the problem so that you can clearly state the Goals and Constraints, then rank the group of Goals and the group of Constraints along a comparative scale of importance of 1 (least) to 9 (most). The rank is based on your analysis of the problem and your educated or gut opinion of the relative importance of each factor.

Next you measure each of your Alternatives against each of the Goals and Constraints, asking yourself the question, How well does this Alternative meet this Goal–Constraint? This question is also answered on a scale of 1 to 9.

E-MAIL   If you have files in a directory other than the one containing
FROM   the Decision Maker, copy them from the Knowledge Builder
DR. FUZZY   directory, from the DOS prompt or in Windows.

The Decision Maker has one more way for you to customize the decision process. Are you an optimist or a pessimist? You rate your own bias (optimism–pessimism level) and enter it into the decision-making process.

Now you're ready to tell the program to decide. The results are displayed on a bar chart and can be translated into a text report. All the elements that went into the decision are included.

Each project is saved in a single file with the extension *.dec*.

The opening screen presents the File menu, for starting a new project or opening an existing one, and a Help menu. Context-sensitive help in hypertext format is always available within the program.

DECIDING WHICH COLLEGE TO ATTEND

If you've been to college, you know that deciding which school to attend is difficult. If college is in your future, you may already have discovered that there are lots of things to consider. Which schools am I qualified to attend? Which ones are in my price range? Do I want a big university or a small college? Do I want to go away from home or remain here? Which schools offer the courses I'm interested in? What kinds of part-time work are available on campus or nearby? What do they do for fun?

What follows is a way of organizing this information for a student-to-be named Pat Press (a relative of Dr. Fuzzy).

Start out by double-clicking on the Fuzzy Decision Maker™ icon, to load the program. Next, click on New in the File menu. A dialog box (Figure 5.1) asks you to name the project. Type in the name *college* and click on OK.

You will automatically be presented with a series of spreadsheet boxes in which you will type in the short names and the fuller descriptions of the Goals, Constraints, and Alternatives. If you wish to skip one of the boxes, simply click on OK. Be sure to return to the box later and fill it in.

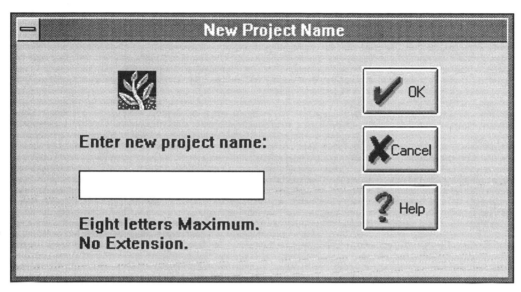

Figure 5.1: Project Name dialog box.

**E-MAIL
FROM
DR. FUZZY**

You can close the program at any time by using the standard Windows close item or by clicking in the upper left-hand corner of the screen.

When all windows are closed, the Decision Maker's screen will display the full program menu and, just below it, the icon toolbar (Figure 5.2). The icons are defined in Table 5.1.

The upper line to the right of the toolbar names the function of the icon under the cursor. The lower line states the active Importances, Satisfactions, and inference method options, as checked in the Names menu's Options box.

To save what you've done, click on the Save icon, or select Save from the File menu.

Figure 5.2: Full Decision Maker menu and toolbar.

TABLE 5.1: Icon Definitions.

Icon	Definition
	Open file
	Save file
	Start new project
	Name goals
	Name constraints
	Name alternatives
	Define importances
	Define satisfactions
	Decision process
	Close windows

E-MAIL FROM DR. FUZZY The Fuzzy Decision Maker™ also allows manual entry of this (and all other) information, so you can choose a specific screen to work on, such as Name Alternatives.

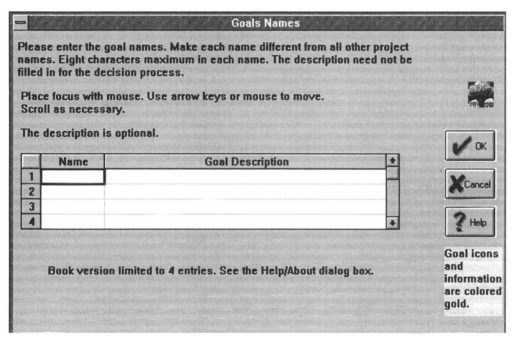

Figure 5.3: Goals Names dialog box.

Naming Your Goals

The first dialog box to be displayed is Goals Names (Figure 5.3).

E-MAIL   For manual presentation of this dialog box, click on the
FROM   Goals icon in the toolbar or select Name Goals from the
DR. FUZZY Names menu.

Pat's college decision involves four Goals. First comes

Prestige—school should have high prestige.

Type *Prestige* in the Names box, press your keyboard's right-arrow key to move the cursor the Description box, and type *school should have high prestige*.

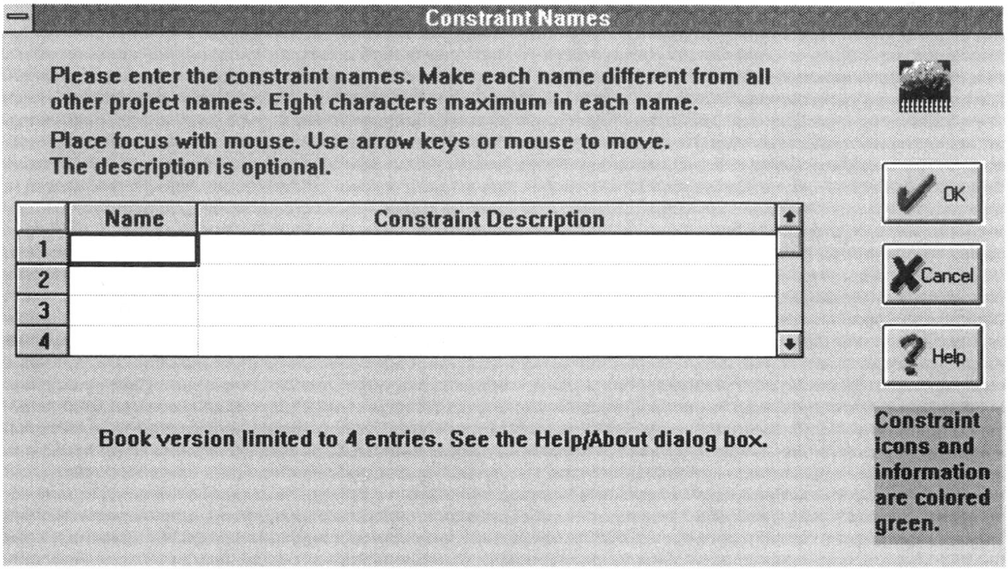

Figure 5.4: Name Constraints dialog box.

Use the four arrow keys to maneuver around the spreadsheet. Each name can have a maximum of eight letters. Each description can be about 40 characters long.

Pat also wants the school to have high professional ratings. This is the second Goal.

Prfslism—high professional ratings.

Where should the school be located? Pat's decided that it would be good to fly far from the nest. The third Goal is

LongDist—a long distance from home.

Finally, Pat's a laid-back person, who prefers that the school have medium living expenses and living standards. The fourth Goal is

Med_buck—med. expenses–living stds.

When you've entered all the Goals, click on OK. The Name Goals box will close and the Name Constraints box (Figure 5.4) will automatically appear.

E-MAIL   For manual presentation of this dialog box, click on the Con-
FROM   straints icon in the toolbar or select Name Constraints from
D R. FUZZY the Names menu.

Name Your Constraints

Pat's goals are limited by some personal preferences and needs. Enter these Constraints just as you did the Goals.

To begin with, Pat is quite interested in languages, so the first Constraint is

Lang_mjr—school must offer lang. major.

Pat is serious about getting an education, but in no way is looking for an ivory tower. So the second Constraint is,

PartySch—the more partying the better.

The last two Constraints involve money. Pat lacks the resources to pay for college completely right now and needs help. So the third and fourth Constraints are

Low_Tuit—tuition should be low.

and

FinanAid—schl shd give financial aid.

When you've finished entering the Constraints, click on OK. The next dialog box is for the Alternatives (Figure 5.5).

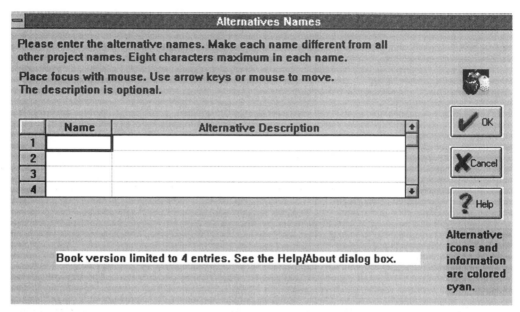

Figure 5.5: Name Alternatives dialog box.

Name Your Alternatives

After long hours at the library pouring over catalogs, Pat has narrowed the
potential-college list to four quite different schools.

E-MAIL   For manual presentation of this dialog box, click on the Al-
FROM   ternatives icon in the toolbar or select Name Alternatives
D R. FUZZY from the Names menu.

The first school is a large private one, named

IvyCvr_U—Ivy Covered University.

Second on the list is

Home_St_U—Home State University.

The third possibility is a small religious school,

St_Al_U—St. Algorithm University.

And fourth is

XtwnColl—Crosstown Community College.

You've now completed entering Pat's basic information. The decision process is about to begin. The next screen lets Pat rank the Goals and Constraints according to their importance (Figure 5.6).

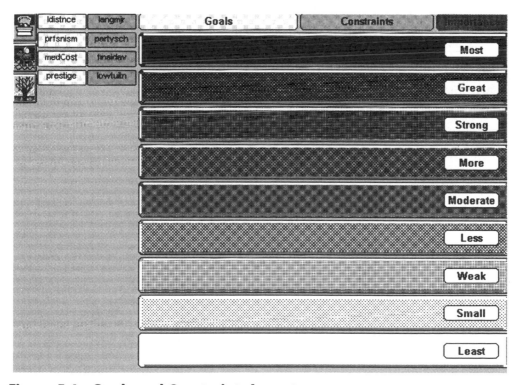

Figure 5.6: Goals and Constraints Importances screen.

E-MAIL   For manual presentation of this dialog box, click on the Im-
FROM   portances icon in the toolbar or select Graphics Importance
DR. FUZZY   from the Importance menu.

Rank the Importances of Your Goals and Constraints

The Importances screen presents a board with nine categories on which Pat
ranks the Goals and Constraints. These are stacked neatly at the left of the
board, the Goals on yellow icons and the Constraints on green ones. There
are also three command icons, which are defined in Table 5.2.

Place each Goal and Constraint by dragging it to the appropriate
board, as listed in Table 5.3. The Importances range from Most, at the top, to
Least, at the bottom. You can rearrange them as much as you want. The
graphics screen provides snap-to guides, so it's easy to align the items on
each board.

When you've finished ranking the Goals and Constraints, save your
work by clicking on the Save icon. Then click on the Close Window icon. The
Importances screen will be replaced by the Satisfactions screen (Figure 5.7).

TABLE 5.2: Importances Screen Icons.

Icon	Definition
	Importances (click on for Help)
	Save file
	Close window

TABLE 5.3: Importances of Goals and Constraints.

Goal or Constraint	Importance Level
Goals:	
Prestige	Small
Prfslism	Great
LongDist	Most
Med_buck	Less
Constraints:	
Lang_mjr	Most
PartySch	Strong
Low_Tuit	Least
FinanAid	More

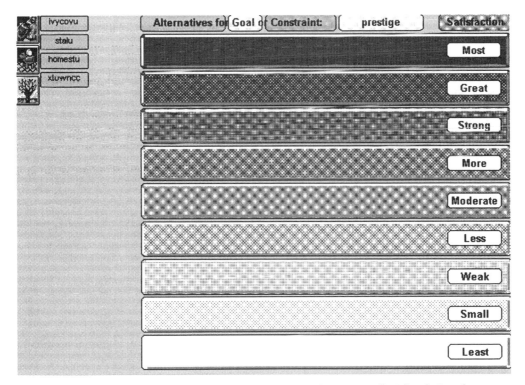

Figure 5.7: Satisfactions screen, Alternatives for an individual Goal or Constraint.

The Importances rankings are also called the *Judgment Scale*. The values are

Ranking	Numeric value
Most	9
Great	8
Strong	7
More	6
Moderate	5
Less	4
Weak	3
Small	2
Least	1

How Well Do the Alternatives Satisfy the Goals?

A Satisfaction screen is set up much like the Importances screen. The cyan-colored Alternatives are stacked to the left of the red-shaded boards. The name of the Goal (yellow) or Constraint (green) is shown above the boards, to the left of the word Satisfactions. The Save and Close Windows icons are at the left of the screen.

A separate screen will appear for each Goal or Constraint, so that the Alternatives can be ranked on each one. After completing the ranking on the first screen, click on the name of the Goal or Constraint and the next Satisfactions screen will automatically be displayed. For instance, after ranking the Alternatives on Prestige screen, click on the word Prestige. The screen for Prfslism will automatically be displayed.

Use the rankings in Table 5.4 to complete Pat's Satisfactions screens for each Goal and Constraint.

When you've completed the series of eight Satisfactions screens (one each for a Goal and Constraint), click on the Save icon and then on the Close Window icon.

The screen will automatically display a graph of the decision process (Figure 5.8). As you can see, Ivy Covered University is the school that best fits Pat's needs, followed closely by Home State University.

If Pat wants to see a text version of the graph, this is the procedure:

TABLE 5.4: Satisfactions Rankings for Each Alternative.

Goal or Constraint	Alternative			
	IvyCvr_U	*Home_St_U*	*St_Al_U*	*XtwnColl*
Goals:				
Prestige	Most	Less	More	Least
Prfslism	More	Less	Most	Least
LongDist	Most	Moderate	Great	Least
Med_buck	More	Most	Moderate	Least
Constraints:				
Lang_mjr	Most	Less	Small	Great
PartySch	Weak	Less	Strong	Moderate
Low_Tuit	Most	Less	More	Weak
FinanAid	Most	Small	Moderate	Weak

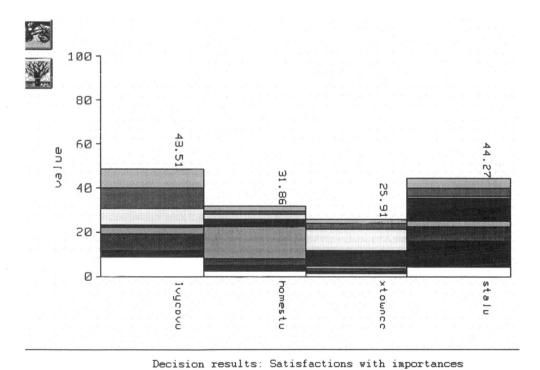

Decision results: Satisfactions with importances

Figure 5.8: The college choices based on Pat's information.

While the graph remains displayed, select Build Report from the File menu. Click on OK in the File Info dialog box.

Then select View from the File menu. Click on the No Conversion button in the next dialog box. In a moment, the text report will be displayed on the screen.

E-MAIL FROM DR. FUZZY

The bar chart can be printed from the screen. Just select Print from the File menu. It lasts on the screen only as long as it's displayed. It's gone when you click the Close Window screen.

If you want to retain it in a file, you can Cut or Copy it to the Clip Board and then Paste it into another document or into a graphics program.

You can build a report only as long as the bar chart is displayed. If you want to build a report at a later time, you must first regenerate a bar chart by clicking on the Decision icon. Then Select Build Report from the File menu.

The report is built and saved as a text (or ASCII) file *college.rep*. It can can later be printed out or pasted into another document. To list the report files in the File Window, select Open from the File menu, click on the File Name window, type *.rep,* and press Enter.

If you want to save the report instead in Write format, click on the Convert file button in the View dialog box.

Need help remembering which color on the bar chart represents? Select Color Palette from the Windows menu. The palette (Figure 5.9) will be displayed on top of the bar chart. It's moveable, so you can place it wherever convenient on the screen.

You can also display the bars as solids. Select Options from the File menu and click to remove the X from the Show Contributing Parts button.

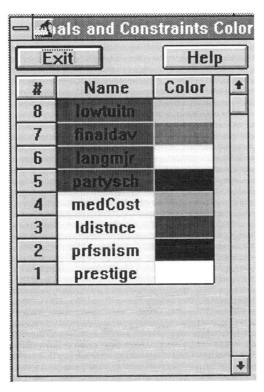

Figure 5.9: Color palette for Goals and Constraints.

REGIONAL TRANSPORTATION SYSTEM

A region's transportation system—its roads, streets, throughways, bus routes, commuter trains, freight trains, ferry boats, cargo ships, harbors, rivers—doesn't exist in isolation. It's part of a dynamic system of people, their homes, the places they go (work, school, leisure areas), how they spend their time, and where and how they earn their money and spend it. If a metropolitan area is a living system, its transportation system is its limbs, arteries, and veins.

These interactions are complex in several ways. First, each of us tries to maximize from the resources that we have and the choices available to us—the land uses and the activity systems that meet our needs. In other words, our lifestyles.

Second, the mix of places where we assemble our lifestyles aren't all in one place. Each person in a metropolitan area has a different mix of housing location, job location, shopping location, school, recreation, and other places to spend resources. How do they do this?

Mobility and transportation are two factors. Either bring the goods and services to the people or bring the people to the goods and services. Like a region's mix of lifestyles, its mobility is complex, or "real murky," as one traffic expert puts it. In other words, fuzzy.

Experts use several kinds of models to understand how transportation systems function and predict the effects on them of the region's future growth, development, and change—land use models and human activity models, based on census and other demographic data; economic models; and air quality models, since air pollution is a product of auto traffic.

E-MAIL FROM DR. FUZZY   This scenario was suggested by Larry Wright, Ebasco Infra-structure Div., Ebasco Corp.

Concepts included in a regional transportation model include System Activity System Distribution, a combination of land-use and socioeconomic data; System Capacity, the number of trips the system was designed to handle; System Availability, the percentage of the region's population who are within easy access of the system (such as one block from a bus line or two miles from a commuter train); Level of Service, which is graded like a school report card, from A (free flow of traffic) to F (gridlock). Transportation experts are always looking for ways to analyze the existing system more realistically and use them in deciding how the system should evolve in the future.

This background is the basis for a project using the Fuzzy Decision Maker™. What changes should be made in the existing system? What social, economic, and governmental constraints limit the kinds of possible changes? What strategies should be used to achieve the desired goals?

Begin by opening a New file in the Fuzzy Decision Maker™ and name the project *Traffic*.

Because this project requires close consideration of the goals–constraints–alternatives relationships, the Decision Maker's Manual system will used for determining Importances and Satisfactions.

To activate the manual feature, Select Options from the Names menu and click on the Importance Manual radio button and on the Satisfaction Manual button. Click on OK. The same screens you've already used for entry of Goals, Constraints, and Alternatives will be presented.

Goals

Four goals have been selected for this example for deciding how to improve a transportation system, shown in Table 5.5.

Constraints

Enter the Constraints listed in Table 5.6.

TABLE 5.5: Transportation System Goals

Name	Description
levserv	Level of svce with some excess capacity
reduauto	Reduce the number of auto trips
redairp	Reduce air pollution
incnonau	Increase nonauto travel

TABLE 5.6: Transportation System Constraints.

Name	Description
actdist	Area socioec and land use patterns
syscap	Capacity of the system
costs	Of maintenance, construction, other
sysaval	System availability—pcent pple served

Alternatives

The Alternatives are strategies based on a list of individual options that can be mixed and matched:

- telecommuting
- van pooling
- work staggering
- reducing the number of single-occupancy vehicles
- car pooling
- constructing new highways and roadways for autos
- increasing public transit
- urban goods scheduling
- constructing new operations facilities to better control the existing system

For instance, you can increase existing highway capacity through traffic operations, such as monitoring traffic and metering entrance to high-use throughways. Or you can do it by affecting the way land uses and activity uses affect the system, for instance, encouraging large companies to stagger their work hours.

If you want to even out the demand on the system, you might want to include work staggering, telecommuting, urban goods scheduling, improvement of mass transit, and car and van pooling. If you want to emphasize increasing the efficiency of the existing system, you might want to construct new operations facilities, as well as promoting van and car pooling, and some of the other options.

Each of the four Alternatives used in this project includes a selection of the options.

TABLE 5.7: Transportation System Alternatives

Name	Description
roadcap	Increase road capacity
masscap	Increase public transit capacity
reddmnd	Reduce and shift the demand
opseff	Increase system operating efficiency

Enter the Alternatives listed in Table 5.7.

So far, you've used the same screens for the Manual method that you did with the Graphic. Now the process actually becomes hands on, as you determine the Importances and Satisfactions.

Importances

To use the Manual method, select Manual Importance from the Importance menu, which displays a dialog box titled Manual Entry of Goals and Constraints Importance (Figure 5.10).

Each possible pairing of Goal and Constraint is presented in the top part of the box, giving you the opportunity to decide which one is more important. For instance, if the Goal–Constraint in the left-hand box is more important, click on radio button 1. If the Goal–Constraint in the right-hand box is more important click on radio button 2. To enter the number of times one or the other is more important, use the up–down arrow keys until the

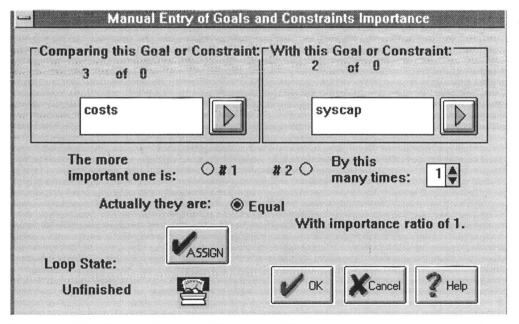

Figure 5.10: Manual Entry of Goals and Constraints Importance dialog box.

desired number—from 1 to 9—appears in the box. Or select the box and enter the number from the keyboard.

If you decide they are equal, click on that radio button. When you have finished with one pair, click on the Assign button to register the values. The next pairing will appear in the boxes. Perform similar routines until you have compared all the pairings.

Notice that the program automatically displays the number of rankings still to be made (above the names) and tells you whether the process is Finished or Unfinished (lower left-hand corner). Be sure to continue until Finished is displayed (a bell will sound). Otherwise, the decision process won't work.

Enter the Importances from the information in Table 5.8, making the italicized goal or constraint the more important in the pairing. When you're done, close the dialog box.

E-MAIL FROM DR. FUZZY

There are two Manual methods of determining the Satisfactions—serial and parallel. In the serial method, the Goal–Constraint and Alternative pairs are presented one at a time. In the parallel method, all the Goals–Constraints are presented together for each Alternative (Manual 1) or else all the Alternatives are presented together for each Goal–Constraint (Manual 2).

To use the serial method, select Manual Serial Goals and Manual Serial Constraints from the Satisfaction menu. Figure 5.11 is an example of a dialog box for this method.

To use the Manual 1 method, select Manual Parallel Goals 1 and Manual Parallel Constraints 1 from the Satisfaction menu. To use the Manual 2 method, select Manual Parallel Goals 2 and Manual Parallel Constraints 2 from the Satisfaction menu.

Satisfactions

The Transportation project will use the Parallel Manual method 1. It presents all the Alternatives for each Goal or Constraint.

TABLE 5.8: Transportation Project Importances.

Goal–Constraint 1	Goal–Constraint 2	Value
levserv	redairp	equal
reduauto	redairp	2 times
levserv	incnonau	4 times
reduauto	incnonau	equal
redairp	incnonau	equal
levserv	actdist	6 times
reduauto	actdist	7 times
redairp	actdist	4 times
incnonau	actdist	8 times
levserv	syscap	2 times
redauto	syscap	3 times
redairp	syscap	equal
incnonau	syscap	5 times
actdist	syscap	2 times
levserv	costs	2 times
reduauto	costs	equal
redairp	costs	3 times
incnonau	costs	2 times
actdist	costs	5 times
syscap	costs	2 times
levserv	sysaval	2 times
redauto	sysaval	equal
redairp	sysaval	2 times
incnonau	sysaval	4 times
actdist	sysaval	equal
syscap	sysaval	2 times
costs	sysaval	2 times
levserv	redauto	4 times

For the Goals' Satisfactions, select Manual Parallel Goals 1 from the Satisfactions menu, displaying the dialog box in Figure 5.12. For the Constraints' Satisfactions, select Manual Parallel Constraints 1 from the Satisfactions menu, displaying the dialog box in Figure 5.13. The Alternatives are

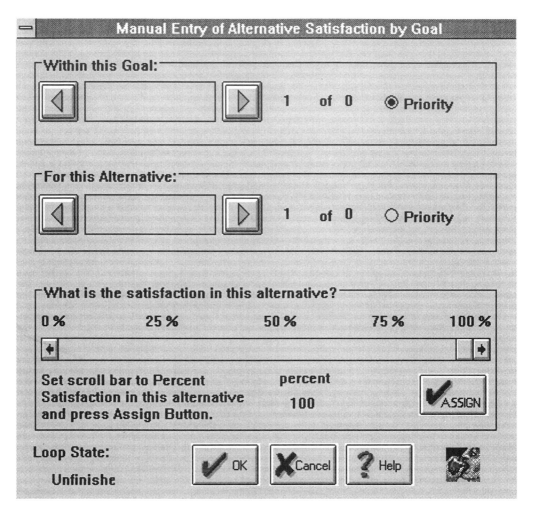

Figure 5.11: Manual Entry of Alternative Satisfaction by Goal dialog box, for serial manual entry.

listed on the spreadsheet, and a the name of a Goal or Constraint is listed in the box.

Enter the information from Table 5.9. When you've completed the first Goals spreadsheet, click on Assign. The next Goals sheet will appear. When you've finished all the Goals Satisfactions, click on OK.

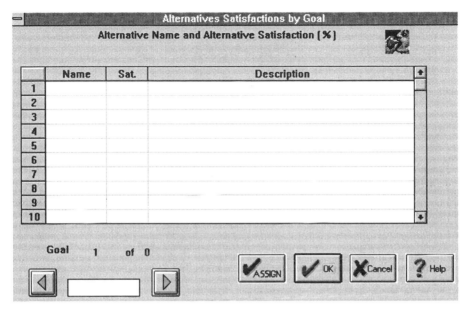

Figure 5.12: Alternatives Satisfactions by Goal dialog box for the Parallel 1 method.

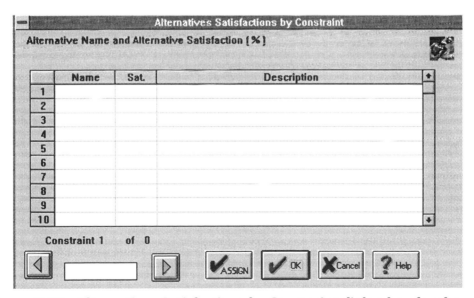

Figure 5.13: Alternatives Satisfactions by Constraint dialog box for the Parallel 1 method.

TABLE 5.9: Percentage Values for Transportation Satisfactions.

Goal or Constraint	Alternative	Percentage
Goals:		
levserv	roadcap	40
	masscap	50
	reddmnd	90
	opseff	80
reduauto	roadcap	0
	masscap	40
	reddmnd	20
	opseff	30
redairp	roadcap	0
	masscap	90
	reddmnd	60
	opseff	50
incnonau	roadcap	0
	masscap	95
	reddmnd	30
	opseff	60
Constraints:		
actdist	roadcap	40
	masscap	60
	reddmnd	50
	opseff	70
syscap	roadcap	50
	masscap	40
	reddmnd	80
	opseff	20
costs	roadcap	95
	masscap	95
	reddmnd	20
	opseff	50
sysaval	roadcap	90
	masscap	80
	reddmnd	20
	opseff	20

For the Constraints' Satisfactions, select Manual Parallel Constraints 1 from the Satisfactions menu and enter the data as you did with the Goals spreadsheets.

When you're done, Save your work.

The Decision Process

Now you're ready to make the decision. Click on the Decision icon and the bar chart will be displayed. Since a governmental agency usually has many different people involved in a decision, you will probably want to consider the decision with various biases. Table 5.10 summarizes the bar charts at biases 0, -50 (medium-pessimistic), and +50 (medium-optimistic).

To change the biases, select Options from the Names menu and change the bias number.

You can also display the results with Satisfactions only, ignoring the Importances. In the Options dialog box, click on the Satisfactions Only box and then click on OK.

MERGING INTERESTS

Many kinds of decisions involve merging the interests of two individuals or groups. Politicians do it all the time. A bipartisan foreign policy in the United States, for instance, might involve the business-oriented policies of the Republicans with the human-rights propensities of the Democrats. When two

TABLE 5.10: Transportation Bar Charts Summary.

	Satisfactions with Importances			Without Importances
	0	-50	+50	
roadcap	36.86	11.13	67.88	39.38
masscap	68.32	48.04	86.58	68.75
reddmnd	51.07	31.61	71.27	46.25
opseff	52.68	37.98	66.67	47.50

businesses decide to merge, they must be sure that differing priorities or points of views don't outweigh the goals and other aspects that they have in common. This is also the case in personal mergers, such as marriage.

Dr. Fuzzy doesn't promise eternal bliss, but the Fuzzy Decision Maker™ can be a useful tool for potential mergees who want to evaluate their chance of success. How? It's a three-step process:

1. The potential partners work out together what the goals, constraints, and alternatives are.
2. Each potential partner uses the Fuzzy Decision Maker™ privately to rank the goals and constraints, assign importances and satisfactions, and have the program reach a decision.
3. The partners then compare their results. How closely do their importances and satisfactions agree? Did the partners' actions result in selection of the same alternative?

Here's an example from real life.

The Scenario

Not wanting to get in trouble with friends and acquaintances, Dr. Fuzzy decided to look at some real lives out of the past—in this case, George and Martha Washington. The lives of the first First Family offer an instructive case study.

The Story So Far

George Washington and Martha Dandridge Custis met in their native Virginia when they were in their mid-twenties, in the year 1758. George was a former surveyor of the uncharted western lands, a decorated war hero (French and Indian War), and a not very rich, but eligible bachelor. Martha was a recent widow, extremely rich, with two small children, and was very likely to get married again soon. George wanted to settle down with someone appropriate—"marry up," if possible—make some money from agriculture, and go into politics. He was carrying the torch for a neighbor, a married woman named Sally Carey Fairfax.

Today's Episode

Can George and Martha find happiness together? Or should they break up and keep looking for someone else. Of course they did decide to marry. Dr. Fuzzy likes to think this is the decision process they used. Compute along with Martha and George in the Fuzzy Decision Maker™.

The Alternatives

In some kinds of decision making, as you've already seen, setting the goals and constraints is the first order of business. Here's a case where determining the alternatives is the best way to begin.

Select New from the Decision Maker's File menu and name this project *marry_yn*. Because George and Martha wanted to enter the Alternatives first, when the Goals screen is automatically presented, simply close it by clicking Cancel.

George and Martha's alternatives are simple and straightforward:

>   marry—Get married to each other.
>   no_marry—Don't get married to each other.

Click on the Name Alternatives icon (or select Names Alternatives from the Names menu), enter the alternatives information, and click on OK.

Next, Martha and George can turn their attention to the goals and constraints.

The Goals

Determining the goals is more difficult, because any merger-related goal will be important to the individual and to *a* potential partner. However, fullest achievement of even a mutually agreed upon goal may not necessarily involve *this* potential partner.

Click on the Name Goals icon and enter the goals information that follows. Martha and George can probably set the first goal quickly:

>   50_years—Be satisfactorily married for many years.

Agriculture was the main economy in colonial Virginia, and George and Martha were both born into the plantation-owning socioeconomic class and both consider it their way of life. So the second goal is

ProfFarm—Operate a profitable plantation.

Both of them are conventional in attitude toward families, and fond of children, so it's likely that another goal will be

HaveKids—Have children with a spouse.

Finally, Martha and George both like to have a good time, so the last goal is

SoclLife—Have an active social life.

When you're done on this screen, click on OK. Be sure to save your file by clicking on the Save icon.

The Constraints

Click on the Name Constraints icon and enter the names, as follows.
    The constraints reflect the potential for conflicts between the partners. Take Martha and George, for instance. First of all

GSoldier—George is a career soldier.

He's newly retired to the civilian sector, but George has a strong sense of duty. So in any new war he'll be likely to get back into uniform and turn away from the country life that Martha loves. Even without a war, a man of action like George might get tired of the country life and want to return to active service. Either case will leave Martha alone to manage the plantation business, as well as bring up the children. Maybe this isn't much different from being a widow.
    Next,

M2kids—Martha has two children from her previous marriage.

Will she have conflicting loyalties and priorities between her new husband and her children?

Then there's

G_Torch—George's old flame, Sally.

Sally and her husband own the plantation next door to George's. Will George be able to focus on marriage with Martha, with the old flame living just down the road?

Finally, there's the wealth differential.

GPorMRch—George is strapped for cash to fix up his plantation, Mt. Vernon, and Martha is very rich.

Martha has inherited several plantations from her husband. In fact, she's the richest woman in Virginia. Will she be able to hack it in more modest surroundings? Will George be jealous of his wife's vast wealth? Will he expect her to subsidize his plantation?

When you've entered the constraints, click on OK and Save the file.

So far, George and Martha have worked together to develop the alternatives, goals, and constraints. Now's the time for them to take these same basics and work on them separately. Martha will do her importances and satisfactions, and set her bias. George will do his.

Perform two Save As operations on the existing program, naming one *G_Marage* and the other *M_Marage*.

```
-------------------------------------------------------------------------
E-MAIL      This is really a role-playing exercise. It works best if you carry
FROM        one character at a time all the way through the process. (Un-
DR. FUZZY   less you enjoy morphing from one personality to another!)
-------------------------------------------------------------------------
```

George's Version

Open *G_Marage.dec* and select Graphics Importance from the Importance menu. George ranked the goals and constraints as listed in Table 5.11.

TABLE 5.11: George's Importances of Goals and Constraints.

Goal or Constraint	Importance
Goal:	
50_years	Moderate
ProfFarm	Most
HaveKids	Least
SoclLife	Less
Constraint:	
GSoldier	Weak
M2Kids	Less
G_Torch	Most
GPorMRch	Least

TABLE 5.12: George's Satisfactions Rankings for the Alternatives.

Goal or Constraint	Marry	No_Marry
Goal:		
50_years	Strong	Least
ProfFarm	Great	Great
HaveKids	Moderate	Least
SoclLife	Strong	Great
Constraint:		
GSoldier	Moderate	Great
M2Kids	Moderate	Least
G_Torch	Less	Strong
GPorMRch	Great	Least

He next decided how much each goal and constraint satisfied the alternatives. George's Satisfactions are listed in Table 5.12.

Finally George determined his Bias setting in the Names menu's Options item. He decided that he was reasonably optimistic and selected a setting of +30.

When he pressed the Decision icon, George's graph looked like the one in Figure 5.14. To create a text report of the decision, select Build Report

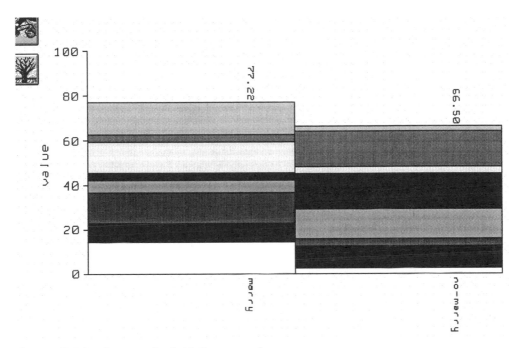

Figure 5.14: George's decision graph.

from the File menu, then view it by Selecting View Report from the File menu (Figure 5.15).

Martha's Version

Meanwhile, Martha was working on her own decision process. She decided to use a different way of representing the graphics, called Graph 2, rather than the default Graph 1. Graph 2 presents a Satisfaction screen for each Alternative. The user then ranks all the Goals and Constraints for each Alternative.

Open *M_Marage.dec* and select Options from the File menu. Next, click on the radio button next to Graph 2 in the Options screen's Satisfactions box and click on OK.

Now click on the Define Importance icon (or choose Graphic Importance from the Importance menu) to display the first Satisfactions screen. As with the Graph 1 method, when you've placed all the Goal–Constraint icons

Fuzzy Decision Maker
Copyright (c) 1994
Fuzzy Systems Engineering

Project Name: G_MARAGE

Goals, Importances and Satisfactions: (9 is Most, 1 is Least)

50-yrs	Be satisfactorily married for many years
5	7,1
ProfFarm	Operate a profitable plantation
9	8,8
HaveKids	Have children with a spouse
1	5,1
SoclLife	Have an active social life
4	7,8

Constraints, Importances and Satisfactions: (9 is Most, 1 is Least)

GSoldier	George is a career soldier
3	5,8

Figure 5.15: Report of George's decision.

on the first screen's board, click on the Alternative name (colored cyan). The next screen will automatically appear. Because there are just two Alternatives, when you're finished the second one, save your values and click on the Close Window icon.

As before, rank the goals and constraints as listed in Table 5.13.

Martha's satisfactions are listed in Table 5.14. Enter those in her program's Satisfactions box.

**E-MAIL
FROM
DR. FUZZY**   The Graph 2 method also allows entry of numerical Satisfactions values in spreadsheet format.

Martha's final task was to set her Bias. Her more confident optimism led her to choose a setting of +50.

TABLE 5.13: Martha's Importances of Goals and Constraints.

Goal or Constraint	Importance
Goal:	
50_years	Most
ProfFarm	Strong
HaveKids	Least
SoclLife	Great
Constraint:	
GSoldier	Most
M2Kids	Least
G_Torch	Strong
GPorMRch	Moderate

TABLE 5.14: Martha's Satisfactions Rankings for the Alternatives

Goal or Constraint	Marry	No_Marry
Goal:		
50_years	Most	Least
ProfFarm	Strong	Small
HaveKids	Moderate	Least
SoclLife	Great	Great
Constraints:		
GSoldier	Less	Least
M2Kids	Moderate	Moderate
G_Torch	More	Least
GPorMRch	Moderate	Small

Martha's decision graph is shown in Figure 5.16 and her text report in Figure 5.17.

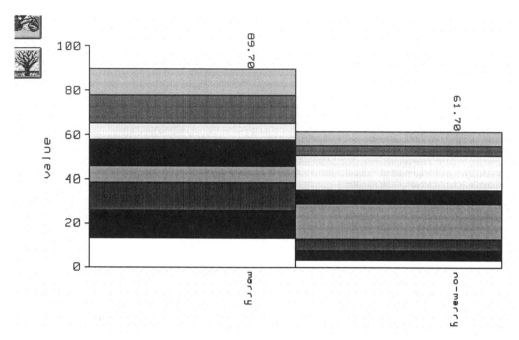

Figure 5.16: Martha's decision graph.

Fuzzy Decision Maker
Copyright (c) 1994
Fuzzy Systems Engineering

Project Name: M_MARAGE

Goals, Importances and Satisfactions: (9 is Most, 1 is Least)

50-yrs   Be satisfactorily married for many years
        9   9,1
ProfFarm   Operate a profitable plantation
        7   7,2
HaveKids   Have children with a spouse
        1   5,1
SoclLife   Have an active social life
        8   8,8

Constraints, Importances and Satisfactions: (9 is Most, 1 is Least)

Figure 5.17: Report of Martha's decision.

Comparing the Two Versions

When Martha and George compared notes, they decided they were pretty compatible and soon got married. As this example shows, using the Fuzzy Decision Maker™ this way lets each potential partner clarify his or her own values and bias and provides a basis for them to jointly evelute areas of agreement and divergence.

E-MAIL FROM DR. FUZZY

George of course did make Mt. Vernon pay off and he did go into Virginia politics. He also went back into uniform in 1776 and left Martha to run the plantation for all the years of the Revolutionary War. They had no children of their own, but raised Martha's children and later adopted two of her young grandchildren after her son died during the war. They lived contentedly ever after, until George's death in 1799.

INSIDE THE FUZZY DECISION MAKER

How does the Fuzzy Decision Maker™ do its work? The Decision menu lets you see some of the inner workings, including the relationships of the Importances and the Satisfactions. These are implementations of the O'Hagan methodology and using the O'Hagan terminology.

Importances

The Importance Levels (which O'Hagan calls Alpha Levels) are the Importances assigned to the Goals and Constraints using the Graphic option (Figure 5.18). For instance, the Goal ProfFarm was placed on the Most board, and so was assigned the value of 9 on the Judgment Scale.

Importance Levels

Exit Help

	Name	Lev.
1	prestige	2
2	prfsnism	8
3	ldistnce	9
4	medCost	4
5	partysch	7
6	langmjr	9
7	finaidav	6
8	lowtuitn	1
9		

Figure 5.18: Importance Levels.

Importance Matrix

Exit

	A	B	C	D	E	F	G	H	I
1	1.00	0.14	0.12	0.33	0.16	0.12	0.20	2.00	
2	7.00	1.00	0.50	5.00	2.00	0.50	3.00	8.00	
3	8.00	2.00	1.00	6.00	3.00	1.00	4.00	9.00	
4	3.00	0.20	0.16	1.00	0.25	0.16	0.33	4.00	
5	6.00	0.50	0.33	4.00	1.00	0.33	2.00	7.00	
6	8.00	2.00	1.00	6.00	3.00	1.00	4.00	9.00	
7	5.00	0.33	0.25	3.00	0.50	0.25	1.00	6.00	
8	0.50	0.12	0.11	0.25	0.14	0.11	0.16	1.00	
9									

Figure 5.19: Importance Matrix.

The fuzzy choices that the user makes in determining the Importances of Goals and Constraints are calculated along with a weighting method developed by Thomas L. Saaty. The results of this calculation—in a notation called *matrix algebra*—are displayed in the Importance Matrix (Figure 5.19).

The final stage in the calculation is of the Importance Factors, which are derived from the matrix values. If you select the satisfactions only, ignoring the Importances, all the Importance Factors are forced to 1.

If you select the Manual option, the program performs a direct calculation of the Importance Factors.

Satisfactions

The Graphic Satisfactions are also calculated in a multistage process. The Satisfaction Levels spreadsheet (Figure 5.20) shows the results of ranking the Goals–Constraints with the Alternatives. Again, Most equals 9 and Least equals 1.

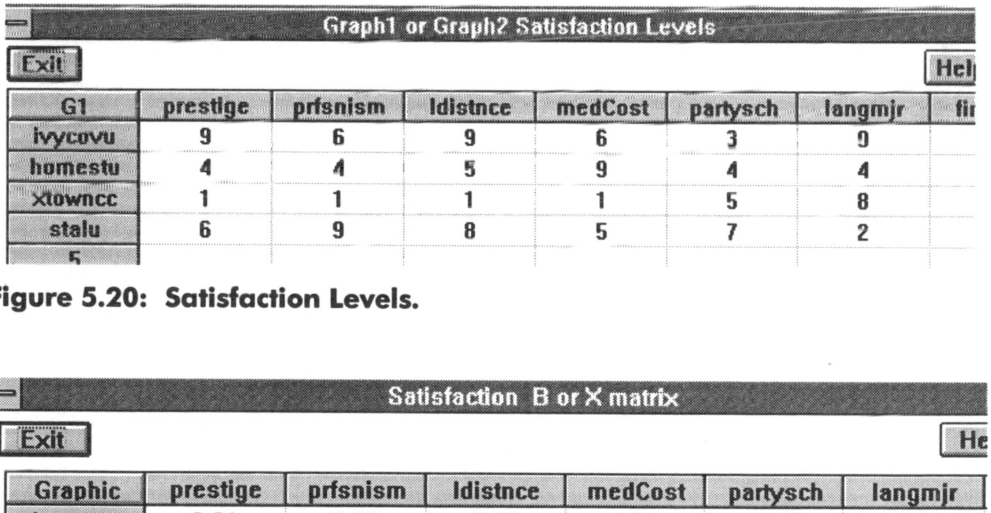

G1	prestige	prfsnism	ldistnce	medCost	partysch	langmjr	fir
ivycovu	9	6	9	6	3	9	
homestu	4	4	5	9	4	4	
xtowncc	1	1	1	1	5	8	
stalu	6	9	8	5	7	2	

Figure 5.20: Satisfaction Levels.

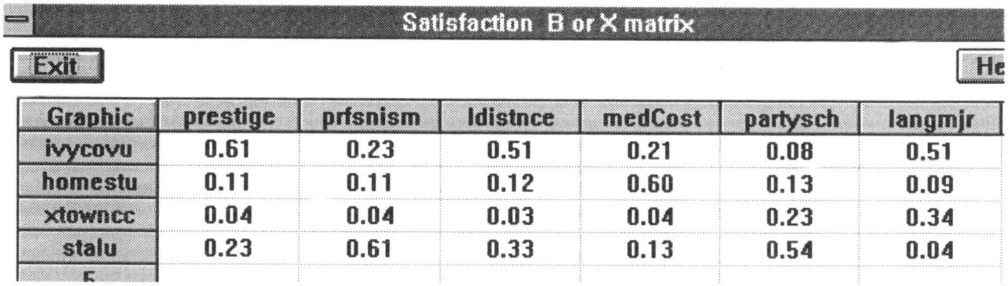

Graphic	prestige	prfsnism	ldistnce	medCost	partysch	langmjr
ivycovu	0.61	0.23	0.51	0.21	0.08	0.51
homestu	0.11	0.11	0.12	0.60	0.13	0.09
xtowncc	0.04	0.04	0.03	0.04	0.23	0.34
stalu	0.23	0.61	0.33	0.13	0.54	0.04

Figure 5.21: BMatrix.

As with the Importances, the Satisfactions are manipulated mathematically, producing the Satisfactions Matrix. Further calculations produce what O'Hagan calls the BMatrix (Figure 5.21). If you use the Manual option, the program calculates the BMatrix directly.

The Decision

The final phase of the Decision process is running the BMatrix plus the Bias through the program's inference engine. As the Option box indicates, two inference methods are available. The *Yager method* (the default) was designed by Ronald R. Yager for fuzzy problems in which goals and constraints are ranked in importance. It's nonlinear in character and involves aggregation operations. It uses the Importance Factor as a exponent for each value in the BMAtrix. The *Perron method* is simpler and linear. It multiplies the Importance Factor by each value in the BMatrix.

E-MAIL FROM DR. FUZZY

You can think of the extreme optimism bias as use of the MAX (fuzzy *Or*) operator for combining sets—a wide-open method designed to incorporate as much as possible. The extreme pessimism bias uses the much more restrictive MIN (fuzzy *And*). The rest of the bias range falls somewhere between MAX and MIN.

In decision making, the aggregation method takes the rankings and combines them using an operator somewhere between MAX and MIN operator.

The book version of the Fuzzy Decision Maker™ uses an aggregation technique developed by Michael O'Hagan called ME-OWA (maximum entropy order weighted average). The commercial version also provides other aggregation techniques.

The BMatrix and the Bias are processed through the inference engine, producing the Decision Factors (Figure 5.22). This box shows the same values that are represented by the bar chart heights.

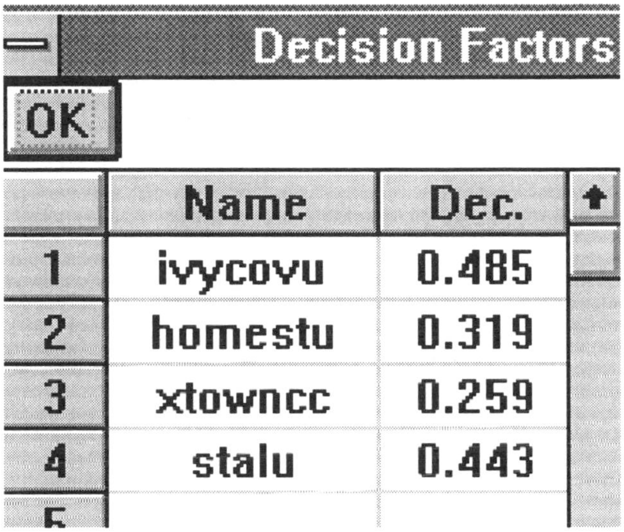

Decision Factors

	Name	Dec.	
1	ivycovu	0.485	
2	homestu	0.319	
3	xtowncc	0.259	
4	stalu	0.443	
5			

OK

Figure 5.22: Decision Factors.

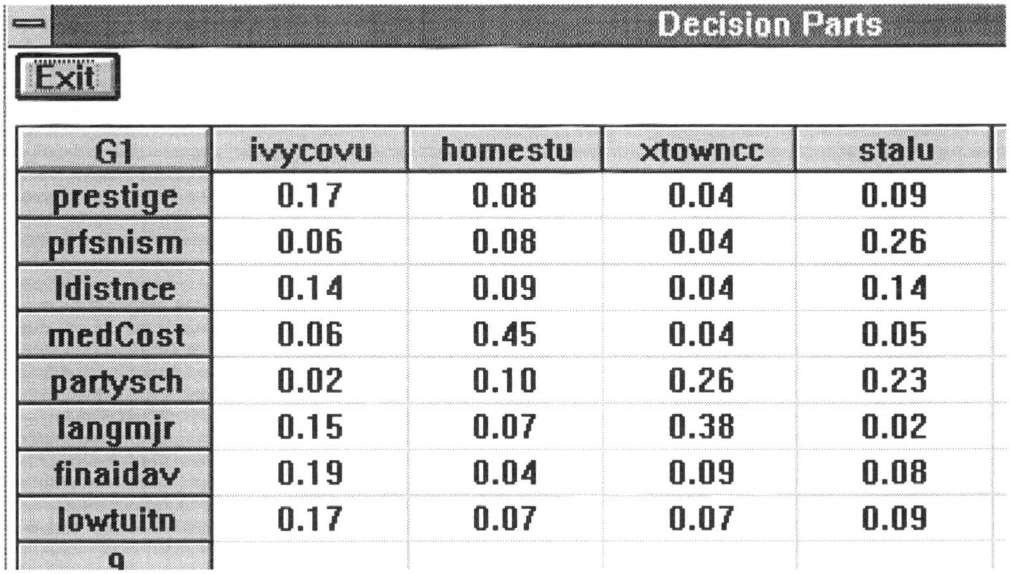

Decision Parts

Exit

G1	ivycovu	homestu	xtowncc	stalu
prestige	0.17	0.08	0.04	0.09
prfsnism	0.06	0.08	0.04	0.26
ldistnce	0.14	0.09	0.04	0.14
medCost	0.06	0.45	0.04	0.05
partysch	0.02	0.10	0.26	0.23
langmjr	0.15	0.07	0.38	0.02
finaidav	0.19	0.04	0.09	0.08
lowtuitn	0.17	0.07	0.07	0.09
q				

Figure 5.23: Decision Pieces.

The Decision Pieces box (Figure 5.23) shows the contributing parts of the Decision Factors. Unlike the colored sections of the bar chart, these values are not normalized.

You've now completed your introduction to fuzzy decision-making. How else can fuzziness be used in the practical world? Chapter 6 provides the third tool, the Fuzzy Thought Amplifier™.

CHAPTER 6

FUZZY THOUGHT AMPLIFIER™ FOR COMPLEX SITUATIONS

In this book so far you've investigated two kinds of fuzzy situations, those that need a person's expertise to control or resolve and those requiring a decision. Other fuzzy situations are also found in the world around us, situations that are both extremely complex and extremely dynamic. The ecosystem of a forest, lake, or prairie is one example. So is a sociopolitical system, such as the Apartheid that until the early 1990s governed South Africa. The effect of bad weather on freeway driving, conflicts in the Middle East, behavior of the stock market, the health care system, international drug trafficking, and even hunger in the human body.

Do these complex and dynamic situations have anything in common? Yes, says Dr. Fuzzy. Each one consists of a group of variable concepts that effect each other. To remain functionally dynamic, the system requires a balance of their interactions. If one of the concepts is greatly changed or removed entirely, the entire system may grind to a halt, flail around endlessly, or become extremely chaotic and useless. To model this form of fuzzy situation, Dr. Kosko invented cognitive maps.

DYNAMIC COMPLEXITIES IN EVERYDAY LIFE

Take your morning commute on the expressway, as originally envisioned by Bart Kosko. In this case, the concepts include the bad weather itself, expressway congestion, auto accidents, frequency of police patrols, your own aversion to risky driving, and your own driving speed. The concepts affect each other, either directly or indirectly and either positively or negatively. For instance, bad weather always leads to congestion of the roadway and it usually leads to auto accidents. The greater is the congestion, the greater the number of accidents, and unfortunately, the greater is the number of accidents, the greater the congestion. Roadway congestion has a strong negative effect on your driving speed—the greater the congestion, the slower your speed. As the frequency of police patrols increases, the number of auto accidents goes down a little. The number of accidents doesn't directly affect your own driving speed, but if the number increases, your own sense of risk aversion will also go up a little. And as your risk aversion goes up, you tend to slow down a little. And so on.

International drug trafficking (a scenario developed by Rod Taber, formerly of the University of Alabama, Huntsville) involves the concepts of drug availability, drug cartels, street gangs, drug usage, profits, corruption, police action, the price of cocaine, and others. Here again, the interdependence can be direct or indirect and positive or negative.

Each of these scenarios can be represented on paper or in a computer as a special type of graph called a *fuzzy cognitive map*. The variable concepts are represented by the nodes, which can also be called *conceptual states*, and the interactions by the edges, or *causal events.*

As the scenarios suggest, in human events the fuzzy cognitive map naturally represents a human way of thinking. Taber has noted that other methods of depicting the cyclic nature of this causal knowledge either ignore the cycles or ignore the importance of real time. The cognitive map allows for both. Because the fuzzy cognitive map organizes dynamic information in such a humanlike way, we call our version of it the Fuzzy Thought Amplifier™.

**E-MAIL
FROM
DR. FUZZY**

A *graph* is a way of representing data. It consists of nodes (also called *vertexes*) and their connectors (called *edges* or *arcs*). In the Fuzzy Thought Amplifer™, nodes are called *conceptual states*.

In some graphs, the connections are directional, meaning that the action is from one node to another and represented by an arrow. This type of graph is called a *directed graph* or *digraph*. In the Fuzzy Thought Amplifer™, the arrows are called *causal events*.

If the type of action is represented as positive or negative, the graph is called *signed*.

A cognitive map is a *signed, directed graph* and can be either crisp or fuzzy.

ORIGINS OF COGNITIVE MAPS

Cognitive maps were formally introduced by Robert Axelrod, a political scientist, in his 1976 book, *Structure of Decision: the Cognitive Maps of Political Elites*. His work represented the map as a crisp matrix.

Crisp Cognitive Maps

The logical structure of crisp cognitive maps allow each state to have a value of either 0 or 1. The connection between two states can have one of three weights (values):

- +1, meaning that the originating or causing state results in an increase in the target or affected state;
- -1, meaning that the causing state results in a decrease in the affected state;
- 0, meaning that the causing state does not change the affected state.

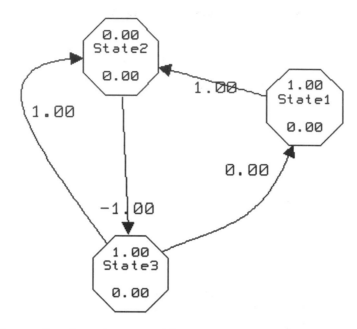

Figure 6.1: Example of a crisp cognitive map.

For instance, the three states (Figure 6.1) in a crisp cognitive map might have these values:

> State 1: 1
> State 2: 0
> State 3: 1

and these connections:

> State 1: — +1 → State 2
> State 2: — −1 → State 3
> State 3: — 0 → State 1
> State 3: — +1 → State 2

The event matrix is shown in Table 6.1.

Because the connection has both direction and a value, it can be considered a vector that changes the value of the target state. This new value

TABLE 6.1: Matrix for an Example Crisp Cognitive Map.

		Affected State	
Causing State	State 1	State 2	State 3
State 1	0	+1	0
State 2	0	0	-1
State 3	0	+1	0

is then enlarged or made smaller ("squashed") through a step function at 0 (greater than 0 equals 1; 0 or less equals 0). This form of limiting function is equivalent to the logistic function with a very large gain.

E-MAIL FROM DR. FUZZY
The cognitive map has many features in common with the artificial neural network. In fact, Kosko refers to a cognitive map as an associative memory, as neural networks are also called.

A networklike use of a fuzzy cognitive map will be presented later in the chapter.

Fuzzy Cognitive Maps

The fuzzy cognitive map, which is the generalization of the crisp cognitive map, was formally introduced by Bart Kosko in 1987. As you might expect, the geometric pattern is identical to the crisp, but states may take on any value over the range {0, 1}. Weights may be limited to the range between +1 and -1 or they may be unlimited. After the connections are made, the new state value is "squashed" through the logistic function. This is the squashing method used throughout the chapter. The "gain" of the function is normally 1.0 though it may vary over anything that is pragmatic.

Any state may also have a feedback loop, allowing it to produce its own causal event. Like other connections, the feedback loop may have a value between +1 and -1.

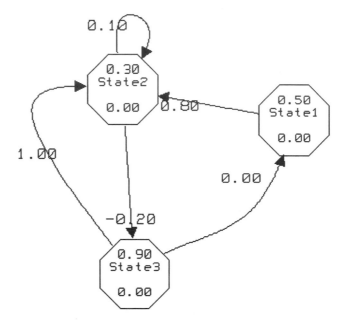

Figure 6.2: A fuzzy cognitive map with three states, one with a feedback loop.

The three states of a fuzzy cognitive map (Figure 6.2) might have these values.

State 1: .5
State 2: .3
State 3: .8

and these connections:

State 1: — +8 → State 2
State 2: — −.2 → State 3
State 2: — +.1 → State 2
State 3: — 0 → State 1
State 3: — +1 → State 2

Its matrix is shown in Table 6.2.

**E-MAIL
FROM
DR. FUZZY**

The states in fuzzy cognitive maps are a type of small computer program called a *state machine*, which was developed more than 50 years ago by Alan Turing, one of the founders of computer science. A state machine, like a mechanical machine, receives something (material, energy, or information) from the outside, uses it ("works"), changes, and perhaps "exports" a product. A state machine may have the ability to change between two states or among a limited or unlimited number of states.

Individual state machines can be linked in various architectures. As a group of *cellular automata* (used in the Fuzzy Knowledge Builder™, for instance), all the state machines are subject to the same set of rules for state changes and outputs and all changes occur more or less simultaneously.

In cognitive maps, state machines are linked in a graph, each one receiving unique inputs from other states, changing as a result, and possibly affecting some or all of the other states. Time is a component of this architecture, because dynamic action continues as long as one state is able to effect a change in another one.

Appendix E describes other architectures for using state machines in fuzzy, time-dependent situations.

Kosko has also demonstrated that, if one fuzzy cognitive map represents the knowledge of one expert, several maps, each representing a different expert, can be combined or superimposed on each other, to produce a consensus representation.

TABLE 6.2: Matrix for an Example Fuzzy Cognitive Map.

Causing State		Affected State		
		State 1	State 2	State 3
State 1	0	+.8	0	
State 2	0	+.1	-.2	
State 3	0	+1	0	

**E-MAIL
FROM
DR. FUZZY** Fuzzy cognitive maps are just beginning to enjoy some recognition outside of the inner circle of developers. To the best of our knowledge, only one person is using them commercially, Dr. Derek Stubbs, of Vicksburg, Michigan, who has developed a stock market analysis program.

Now it's time to examine the basics of cognitive maps by working with the Fuzzy Thought Amplifier™.

FUZZY THOUGHT AMPLIFIER™

The book version of this program allows you to take a real-life scenario and design a cognitive map for it with as many as six states and events going from each state to each of the other ones. (The commercial version provides a maximum of 25 states per map.) Any state can also have a feedback event loop, providing the state's own output as input. In addition, any specific state can have a "bias," a phantom state whose event acts only on the specific state. In addition, you can create a crisp map as well as a fuzzy one.

With this tool, you'll learn how to construct a fuzzy cognitive map from a scenario, customize it for graphic purposes, and create a history of its dynamic behavior. The Fuzzy Thought Amplifier™ runs in two basic modes, normal and trained.

Normal Operation

For normal operation, you can set the initial value of each state and the initial weight for each event, then run the map to see how it behaves dynamically. You'll start by examining the three possible dynamic endings of cognitive maps—stability, oscillation, or chaos. In a scenario-based map, the way it concludes after running may give you some insight into the real world situation it's modeled on.

Next, you'll build two maps that explain the present by reconstructing the past. In one, an animal exhales carbon dioxide, a green plant uses the carbon dioxide and gives off oxygen, which the animal then breathes. You can run the map, and then make it more complex by adding bias and new states.

The second scenario is more complex—an attempt to understand how a national health care system works. (Everyone else is working on it, so Dr. Fuzzy decided to join in.)

"Trained" Operation

The Fuzzy Thought Amplifier™ also provides a neural network-like training function. For this, you assign conditional beginning weights, run sets of historical state values through the map, freeze the "trained" event weights, add new state information, and run the map to predict the future.

Finally, you'll use the past to create and "train" a thought amplifier that will predict the future, in this case the behavior of a stock market.

E-MAIL FROM DR. FUZZY What's the difference between normal running of a cognitive map and training one? Here's a simple rule: Running a map uses fixed weights to change state activations. Training uses sets of state values to change ("train") causal weights until they're correct.

Dr. Fuzzy suggests beginning by examining some very simple, but dynamic cognitive maps.

SIMPLE FUZZY THOUGHT AMPLIFIERS™

To see how simple but dynamic cognitive maps work, open the Fuzzy Thought Amplifier™. From the opening screen, select Open from the File menu, then double click on the file *example.fcm*. The screen now displays three simple cognitive maps (Figure 6.3).

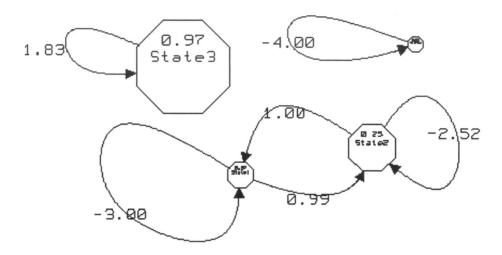

Figure 6.3: Simple cognitive maps.

The icons in the toolbar above the map window have the functions listed in Table 6.3. The upper microhelp line names the icon being pointed to with the cursor.

The lower microhelp line displays the squashing function (or gain) that determines the high and low values of a cycle and can affect the map's operation. The higher the gain is, the more exaggerated the cycle. You can see a graph of the squashing function and change its value (Figure 6.4), if desired, by selecting Squasher from the Run menu.

The three simple cognitive maps were put in the same file because of the limitations of disk capacity. Dr. Fuzzy wants you to examine each one individually, so you can begin learning your way around the Thought Amplifier by separating the maps, starting with one that is stable or static.

TABLE 6.3: Icon Definitions.

Icon	Definition
	Open file
	Save file
	Add a new state
	Delete a state
	Display the list of states
	Name a state
	Add a new event
	Delete an event
	Name an event
	Display the event matrix
	Run one map cycle (step forward)

Stable Map

Start by saving this file under a different name. Select Save As from the File menu, then in the box above the filename window delete the asterisk and type in the name *exmpstat* and press Return.

The next step is to delete the states that aren't required for the stable map—States 1, 2, and 4. Begin by clicking on the Delete State icon in the icon

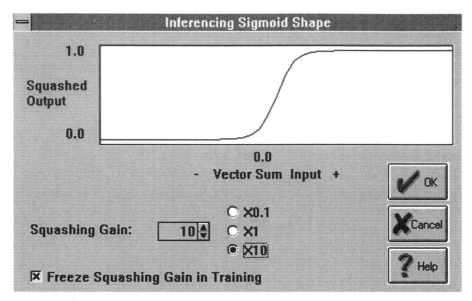

Figure 6.4: Inferencing Sigmoid Shape (squashing function) dialog box.

toolbar. A dialog box will tell you to click on the state that you want to delete. Click OK on this dialog box.

Now click on State 1. It and its causal events will be wiped from the screen.

Perform the same operation for States 2 and 4. When you've finished, only State 3 will remain.

Save the file by clicking on the Save icon.

Now you're ready to see whether the map is dynamic. Click once on the Step Forward icon. You'll see that the state hasn't changed. The value remains at .97. Click on Step Forward again to try another cycle. Again nothing happens. No matter how many times you perform this operation, the state value stays the same, even though it is sending itself a feedback message. The map was created stable and does not change.

In fact, it's so stable that even when you change the value of the state and Step Forward, it returns to the same stable state.

To make the state values easier to see, you can display a box containing a list of state characteristics. Click on the List States icon. When the box is displayed, use your mouse to drag it to a blank place on the screen where it won't obstruct the map. Now you're ready to observe the dynamic oscillations of the state. Note that the initial state value is .97 on the state itself and 97 in the List Box. The two values are really the same, with the decimal point changed in the List Box for computational purposes.

Double-click on the value in the List Box and change it to 80. Now click on the Assign button to register it in the map. Click on the Step Forward icon three times, and note that by the second step the state value will return to 97. It will remain there, no matter how many more steps you take.

When you're finished, close this file by selecting Close from the File menu.

Oscillation

The next map to examine isn't stable. Instead, it ends up toggling between two values as successive cycles are run.

Open the example.fcm file again, this time saving it as *exmposcl* This time, delete States 1, 2, and 3, so that only State 4 remains on the screen. Save the file.

Notice how small State 4 is—so small that you can't read the information it contains. There's a reason for this, a program option called *dynamic sizing*. When the value of a state is small, you can have the state itself displayed in very small size. As the value increases, the size of the state also increases.

Display the List Box to make the numbers easier to read.

E-MAIL FROM DR. FUZZY   You can graphically represent the value of states and the weight of events in several dynamic ways—size; the color characteristics of hue, saturation, or luminance; or a combination of size and color.

Click on the Step icon, and watch how the state's value changes to .44. As this happens, it's reflected in the dramatic increase in the state's size. Click on the Step icon several more times, and observe how the state changes from small to large and back to small again, as its value oscillates to .03, then .44, back to .03, and so on.

Will the oscillation occur if you change the value of the state? Try it and see.

In the List Box, change the state's value from 3 to 10 and click on Assign. Now perform a series of steps. It'll take a while, but beginning on the 10th step, you'll see that the value begins an oscillation between 3 and 44.

If you wish, try it again with a state value of your own choosing. When you're finished, close the file.

Chaos

Once more open example.fcm, this time saving it as *exmpchao* and then delete states 3 and 4. Save the file.

This map is designed to be chaotic, meaning the states will have different values with each cycle. Displaying the List box will help you see the differences, but you may want to compile a complete record of the cognitive map's dynamic behavior by setting up a history file.

E-MAIL FROM DR. FUZZY

To compile a record, you must set up a history file before you start running the cycles.

First, select Initialize History from the Run menu. After you run the cycles, you can see the record by selecting View History from the Run menu. This creates a text file that can be read while the file is on the screen. Simply click on the dialog box's No Conversion button. The values (to three decimal places) are listed in the report.

This history file is saved with the file name and the extension *.hst*. You can format and print it from any word processing program or load it into another document.

Now begin running a series of cycles. As you'll see, the state values will be those shown in Table 6.4. If you continue beyond seven steps, the values will continue to be chaotic, and no pattern will ever appear.

To verify the generally chaotic character of this map, you can change the value of one of the states and both of them, then run the map. For instance,

TABLE 6.4: Chaotic State with the Original Values.

	State 1	State 2
Original value	.07	.25
Step 1	.52	.44
Step 2	.10	.12
Step 3	.41	.56
Step 4	.20	.08
Step 5	.26	.52
Step 7	.38	.13

change the value of State 1 from 7 to 50, assign it, and perform a series of steps. As you'll see, no pattern emerges. The conclusion is still chaotic.

Now return State 1 to its original value of 7. This time change State 2 from 25 to 70, assign it, and step. Again, chaos.

Finally, change both values—State 1 to 80 and State 2 to 40. Assign the new values, then Step Forward. The result is chaos, once again. When you're finished, close this file.

A fourth kind of simple fuzzy cognitive map process is called a limit cycle. One may arise as an emergent global phenomenon, like "the wave" at a sporting event, or it may arise from round-off effects in the inference process. Limit cycles are supported in the commercial version of the product.

**E-MAIL
FROM
DR. FUZZY**
In chaos theory, this map is what is called a *single-point at-tractor.*

The fuzzy cognitive maps you've worked with so far have been operating in normal mode. Now you'll see how a map can be trained with past data so that it can predict the future—or so they say.

Now that you've seen the basic operation of dynamic fuzzy cognitive maps, Dr. Fuzzy will show you how to create one from scratch—really—by deriving a map from the natural environmental interaction between an oxygen-breathing cat and a carbon dioxide-breathing plant. The good doctor calls it CatPlant.

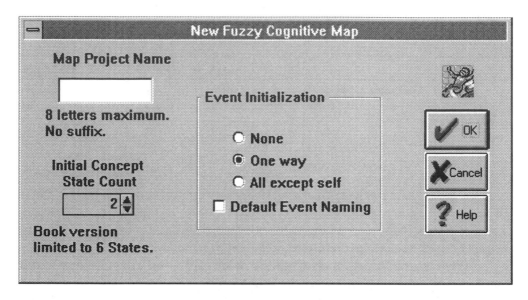

Figure 6.5: New Fuzzy Cognitive Map dialog box.

CATPLANT

The earth's ecological history is greatly shaped by the way plants and animals use two atmospheric gases, oxygen (O_2) and carbon dioxide (CO_2). Animals breathe in oxygen and exhale carbon dioxide, while plants take in carbon dioxide and give off oxygen. Dr. Fuzzy has used this interaction as the basis for a two-state fuzzy cognitive map.

To begin a new map, select New from the File menu. In the New Fuzzy Cognitive Map dialog box that appears (Figure 6.5), place the cursor in the Map Project Name box (top left) and type in *catplant*. Below this is the Concept State Count window. Set this at 2 by clicking on the up–down arrows or typing in the number.

In the Event Initialization box, click on the *None* radio button. When you're finished, click on the dialog box's OK.

You're now ready to name and determine the values of the states.

TABLE 6.5: CatPlant States.

Name	Description
cat	Cat that gives off carbon dioxide
plant	Plant that gives off oxygen

Naming and Defining the States

Begin by selecting Descriptions from the State menu. Type in the name and description of each state, as shown in Table 6.5. When you're finished, click on OK. Now save your work by clicking on the Save icon.

Notice that on the map, each state has already been given the default value of .50. This means that the state begins with 50% of its potential, which is fine for beginning CatPlant.

The next step is to create two events.

Creating Events

To begin, select Add from the Event menu. The Create Event dialog box asks you to click on the starting state, on up to six intermediate points, and the ending state. This is similar to the process of creating a line in a drawing program. Click on OK.

Dr. Fuzzy likes curved event arrows for CatPlant. To create the Cat-to-Plant event, first click on the Cat state. Next click on two or three points in a right-hand arc toward the Plant state. Finally, click on the Plant state itself. A nicely curved event will appear, with the arrowhead at the Plant state.

Now perform a similar operation from Plant to Cat, with the event arrow arcing in the opposite direction. Be sure to save your work.

Event Values and Names

Dr. Fuzzy has determined that the Cat's carbon dioxide exhalation affects the Plant to the extent of 55%, while Plant's oxygen release affects Cat to the extent of 35%. These will be the weights (or values) of the events.

To add these values to the map and display them, select Options from the Run menu (Figure 6.6. Click on the Show Event Names box, and then click on OK.

Defining the Cat-to-Plant Event

To define the Cat-to-Plant event, select Name from the Event menu. The Name and Weight Causal Event dialog box (Figure 6.7) will appear on the screen. In the Causal Event Name box, type in CO_2. In the Causal Event Weight box, insert the value *55*. (For calculation purposes, values are multiplied by 100. The program will automatically convert it to .55.) Now click on OK.

The Name Event dialog box appears, telling you to click on the arrowhead of the event to be named. Click OK on this box, then click on the arrowhead at the Plant state.

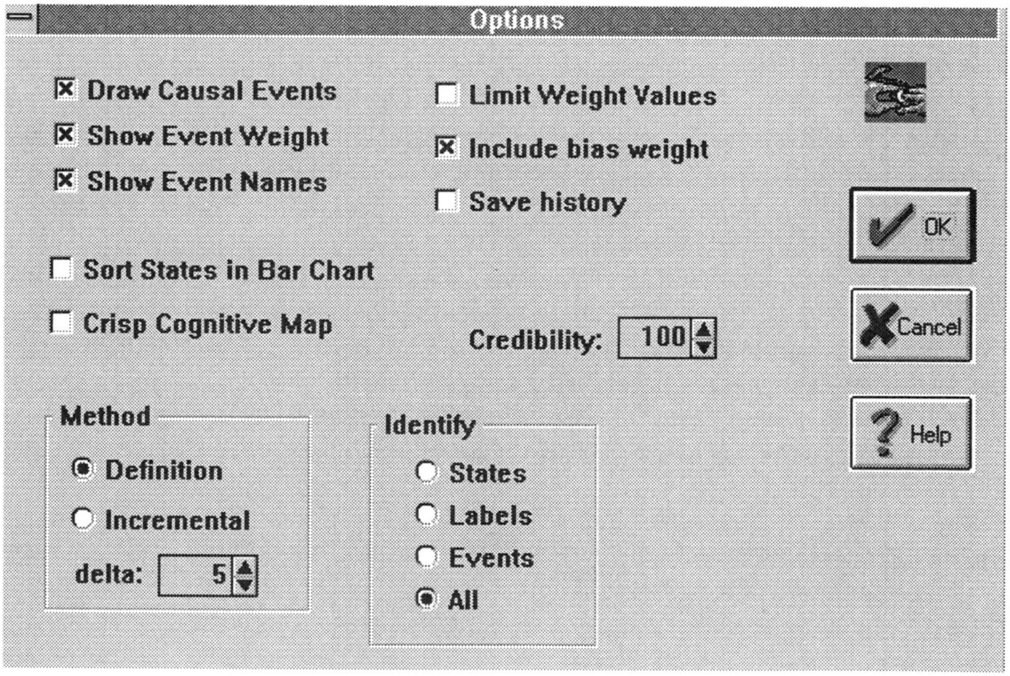

Figure 6.6: Options dialog box.

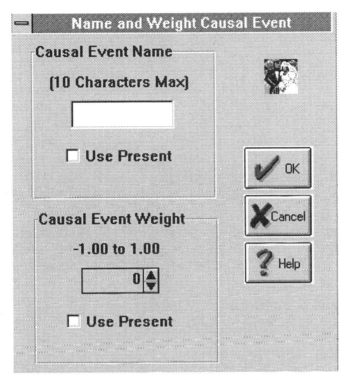

Figure 6.7: Name and Weight Causal Event dialog box.

An Event Move dialog box will then appear, instructing you to click on the map location where you want the event name and value to appear. Click on OK. Then click on a blank space near the arrowhead. The value will appear.

Defining the Plant-to-Cat Event

Perform the same procedure for the Plant-to-Cat event, using the Event Name *O2* and the Event Weight *35*.

Place the information near the Plant-to-Cat arrowhead. Your map should now resemble the one in Figure 6.8. Be sure to save your work.

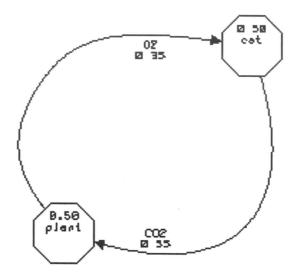

Figure 6.8: Completed two-state CatPlant map.

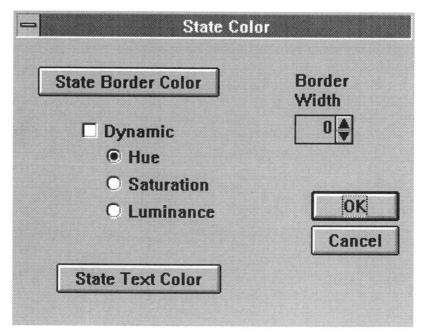

Figure 6.9: State Color dialog box.

You can now start running your map, but first, the Dynamic Dr. Fuzzy suggests that you make the states and events in your map more dynamic by adding color.

Adding Dynamic Graphics

Just as the states in the simple example maps had dynamic sizes, states and events can also have dynamic coloring linked to their values. (Dynamic color and size can be combined, as well.) You can select a color for the state borders and text, and also the width of the borders, so the color is as prominent as you desire.

Colors for States

To add dynamic color to the CatPlant states, select Color from the State menu. Click on the Dynamic button in the resulting dialog box (Figure 6.9).
You can choose one of three color characteristics:

- *hue* is the spectrum color,
- *saturation* is the degree of color, for instance from pale blue to navy blue, or
- *luminance* is the "shininess" of the color, from dull to glossy.

Hue is the default and will be used here.
Now click on the State Border Color button, displaying the color palette. The actual color possibilities that you have depends on the capabilities of your computer and monitor. The color you choose will be assigned to the state value of 100. The Windows environment then assigns other colors to the various other values. Dr. Fuzzy wants to select Red as the state color. Click on the red palette sample, and notice that a black border now appears around it, showing that it has been selected. Now click on OK.
You can also select a border size. To make the color more prominent, change this value to *12*. Now click on OK.

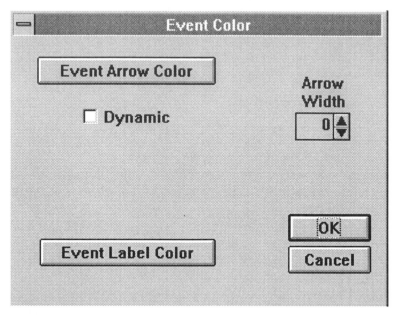

Figure 6.10: Event Color dialog box.

Colors for Events

In a similar process, you can make the Event colors dynamic, too, if your Windows driver supports higher gradations in color. Select Color from the Event menu and enter values in the Event Color dialog box (Figure 6.10).

In the Color dialog box, click on the Dynamic box. Make the arrow wider by changing the width value to 6.

Click on the Event Label button, then select a color, such as orange, from the palette and click on OK. Finally, click on OK in the Event Color dialog box. Be sure to save the file.

One last thing before running the map: Be sure to initialize the history function (from the Run menu), so a record of the cycles will be saved.

Running Cycles

At last, you can run the cycles of your map. Display the List box if you wish.

From beginning values of 50 for Cat and Plant, click on the Step Forward icon. The values will now be 54 and 57. When you step again, they will change to 55 and 58. Step once more and see what happens—the values remain 55 and 58, showing that the map has stabilized.

You can expand CatPlant's complexity by adding a bias.

Adding Bias

A bias is a dynamic, ongoing, but unchanging causal event to a particular state. For example, the Cat uses the ground as its sandbox, releasing nutrients that flow to the Plant. Dr. Fuzzy has decided to give Plant a bias weight of .39. Enter this value in the List box as *39* (Figure 6.11). To register it, click on Assign.

To have the bias value appear on the map, select Options from the Run menu and click on the box labeled *Include bias weight*, then click on OK. Save your work.

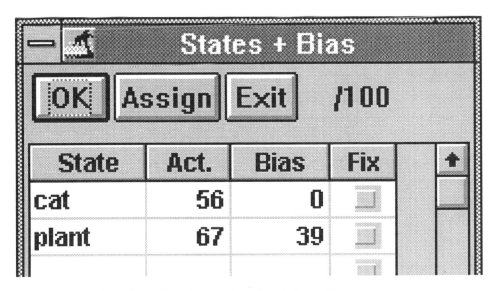

Figure 6.11: CatPlant List box, showing Plant bias.

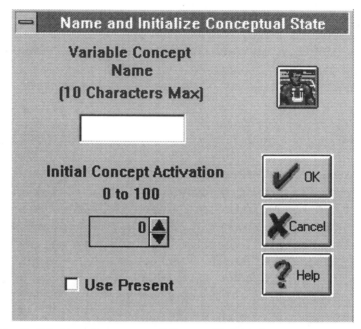

Figure 6.12: Initialize Conceptual State dialog box.

Running Cycles with the Added Bias

From your current state values of 55 for Cat and 58 for Plant, press the Step Forward icon. The values now read 55 and 67. Step again, registering 56 and 67. When you step one more time, you'll see that these are stable values.

Want to make CatPlant still more complex? You can add two more states.

Adding Additional States

Dr. Fuzzy believes in good nutrition for mammals, green plants, and other living things. One way to assure this in CatPlant is by supplying food. For instance, you can add a state named Feed and link it by an event to Cat. You can also add a state named Fertilizer and link it to Plant.

Dr. Fuzzy begins by adding a state near the Cat state, then naming it Feed, representing cat food.

Creating and Naming a State

To add a state, click on the Add State icon. The resulting dialog box will tell you to click on the location of the new state. Click on OK, then click on a blank spot near the Cat state.

Now click on the Name State icon, displaying the Name and Initialize Conceptual State dialog box (Figure 6.12).

In the Name box, type *Catfood* then give it an activation value of *100*. Click on OK.

Repeat this process and create another new state near Plant. Name it *Fertilizer* and give it a value of *100*. Now click on OK.

Creating and Labeling an Event

The next step is creating an event between each new state and its target state. Click on the Add Event icon. Follow the instructions in the Create Event dialog box, clicking on OK and creating the event path. Click first on the Feed state, then on a few points along the way to the Cat state, and finally on Cat. The arrow will appear on the map, its arrowhead near Cat.

Now click on the Name Event icon. In the Causal Event Name box, type *Feed* and give it a Causal Event weight of *31*. Click on OK. Another dialog box will tell you to click on the arrowhead to be named. After you click on the arrowhead near Cat, a dialog box will ask you where the label should be placed. As before, click on a place near the event arrowhead, and the label will appear.

Repeat this sequence to create a Cat-to-Plant event. Name it *Nourish* and give it a weight of *59*. Save your work.

Running the Augmented CatPlant

When you step the cycles, you will see one change to values of 64 for Cat and 72 for Plant and that they are stable.

You can experiment with this map, adjusting initial state values, event weights, and the bias, to see how they affect the map's dynamics. Because you have a historical record of your actions, you can re-create any stage in the map's operations and experiment from that time forward, as well.

When you're finished, close the file.

Bart Kosko has said that you can take any well-written newspaper or magazine article and translate its substance into a fuzzy cognitive map. He based one of his most famous maps, of South Africa's former apartheid system, on a newspaper article by the syndicated columnist Walter E. Williams. This map uses nine conceptual states. Most real-world scenarios require at least that number.

However, it is possible to take a first step toward modeling a situation with the six states available in this version of the Fuzzy Thought Amplifier™. As an example, Dr. Fuzzy shows how to begin considering the United States's health care system.

HEALTH CARE SYSTEM

In the mid-1990s, vast amounts of material were presented to the American public on the subject of health care. With just a few of the many published articles, you can quickly compile a long list of states and ponder the events that connect them. What follows is a first-stage consideration of the problem from one point of view, that of a perhaps typical individual consumer. It's not meant to be the last word, or even the next word on the subject, but it shows how a fuzzy cognitive map can help clarify someone's thinking about a real-world issue.

Start with a long list of possible states, including how the individual person views the state of his or her health care; the nation's level of health; how much actual health care is dispensed for the money spent; how much of the health care money is spent on things other than actual delivery of health services, such as the insurance industry, legal fees, and bureaucracy; whether there are enough physicians to meet the population's needs; and the relative financial contributions to the health care system by individuals, employers, state government, and local government.

The States

Dr. Fuzzy combined some considerations, weeded out others, and finished with six that represented major health care issues and met the criteria for being conceptual states—states they can be changed by means of causal events received from other states in the cognitive map and, in turn, are capable of affecting themselves and other conceptual states through outgoing causal events. The conceptual states decided on are described in Table 6.6.

Naming the States

Open a new file in the Fuzzy Thought Amplifier™ and, in the new project dialog box, name it *hlthcare*. The map requires *six* states. Also, in the Event Initialization box, click on All.

Select Descriptions from the State menu, then enter the names and descriptions listed in Table 6.6.

You'll be doing some rearranging of the states on the map to accommodate the causal events. To make rearranging easier and redrawing of the screen quicker, select Options from the Run menu and make sure that the box for Draw Causal Events is not checked. Click OK. (You can turn the function on after the states are arranged.)

Determining State Values

Each state's value must fall within the range from 0 to 1 (or 0 to 100 in the List box). First, the map builder should decide what each state's values represent. Dr. Fuzzy decided on these criteria for the six states in this map:

- *myhealth* is measured by how cared-for by the system the individual feels, on a scale from 0 (poorly cared-for) to 1 (well

TABLE 6.6: Health Care States and Descriptions.

Name	Description
myhealth	One's own perceived health care coverage
pophlth	Health of the population
efficsys	Efficiency of the system
rspnstim	Response time when one needs care
nonmed	Paramedical infrastructure
noofdocs	No. of doctors per unit of population

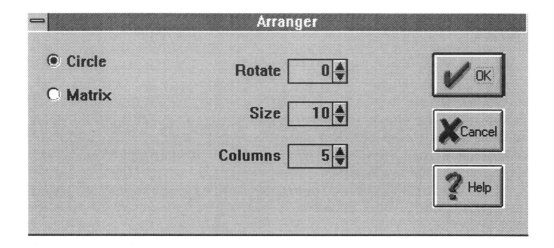

Figure 6.13: The Arranger.

cared-for). In Dr. Fuzzy's case, the value is *.1* (or *1* in the List box).

- *pophlth* is an estimation of the overall health of the public, based on such factors as longevity, infant mortality, the sick homeless, and other standard measurements of health. This state received a value of *.7* (or *7*).
- *efficsys*, an overall evaluation of how much care someone receives for the money involved, received a value of *.3* (or *3*).
- *rspnstim* measures the length of time between the need for care and receiving it, with short = 1 and long = 0. The assigned value here is *.5* (or *5*).
- *nonmed* measures the size of the health care system's paramedical infrastructure, with large = 1 and small = 0. The value assigned is *.4* (or *4*).
- *noofdocs* is a measure of the number of doctors per unit population, based on perception of the situation in Dr. Fuzzy's

own stamping grounds, Southern California. The assigned value is .7 (or 7).

These values can be entered in the List box. Be sure to save the file at this point.

Rearranging the States

The states are arranged on the map in circular formation, which is the default.

The health care map is constructed from the point of view of the individual, so the states should be rearranged with the *myhealth* state surrounded by the other five states, as shown in Figure 6.15.

To move a state, place the cursor within it and hold down the left mouse button until the cursor changes to the "drag" rectangle. Then drag the state to the desired position.

When the states are rearranged, save the file.

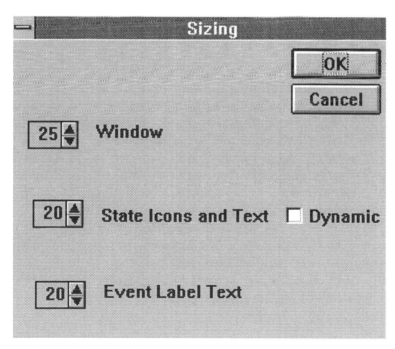

Figure 6.14: Sizing dialog box.

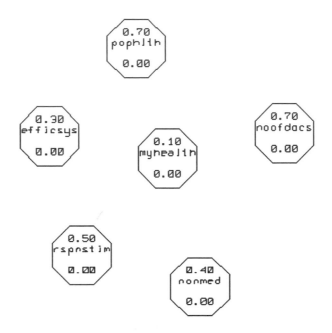

Figure 6.15: State arrangement in the health care cognitive map.

**E-MAIL
FROM
DR. FUZZY**

You can specify the placement of the states or change it through the Arranger. Select Arranger from the Wizard menu. As the box shows (Figure 6.13), states can be arranged in a circle (the default) or in a matrix. You can rotate their positions by specifying a value between 0 and 90. You can also change the size of the map itself—the amount of space between states—by selecting a value between 0 and 100. Or you can arrange the states in from one to six columns.

You can even change the size of the map window itself by selecting Sizing from the Window menu (Figure 6.14). You can also choose the size of the states and their text, make them dynamic, and choose the size of the event labels.

The Events

Now that the states are placed, you can turn on the Draw Causal Events in the Run menu's Options box. Although you won't be using all of the events drawn, in this map it's easier to turn off the unneeded ones than to individually add the 20 events the map requires.

Deleting Events

You can delete an event by clicking on the Delete Event icon. The resulting dialog box will direct you to click on the arrowhead that you want removed. Delete the arrowheads for the events listed in Table 6.7. When you're finished, save the file.

Define Events

The weight of an event is the degree to which one state affects another one, either negatively or positively. For example, the size of the non-medical

TABLE 6.7: Events to be Removed from Health Care Map.

Causing State	Affected (arrowhead) State
myhealth	efficsys
myhealth	rspnstim
myhealth	nonmed
myhealth	noofdocs
pophlth	noofdocs
rspnstim	nonmed
rspnstim	noofdocs
nonmed	pophlth
nonmed	rspnstim
nonmed	noofdocs
noofdocs	nonmed

TABLE 6.8: Healthcare Event Weights.

Causing State	Affected (arrowhead) State	Weight
myhealth	pophlth	30
pophlth	myhealth	70
pophlth	efficsys	10
pophlth	rspnstime	80
pophlth	nonmed	-50
efficsys	myhealth	60
efficsys	rspnstim	90
efficsys	nonmed	-50
efficsys	noofdocs	-20
rspnstim	myhealth	-50
rspnstim	pophlth	80
rspnstim	efficsys	30
nonmed	myhealth	-70
nonmed	efficsys	-70
noofdocs	myhealth	80
noofdocs	pophlth	30
noofdocs	rspnstim	60

infrastructure (State 5) might negatively affect the individual's perceived health care coverage (State 1) with a weight of .7 (or 70).

Dr. Fuzzy has decided that the event weights for this map as shown in Table 6.8. Because there are so many states, the easiest way to enter them in the map is through the Event Matrix.

Click on the Event Matrix icon, which displays the map's matrix (Figure 6.16). Notice that the state names are listed down the left-hand side and across the top. The names on the left-hand side are the causing states; those across the top are the affected states.

The map's events are indicated by a red check mark. Fill in the blank cell below each of these events with the values in Table 6.8. When you're finished, click on Assign. The finished matrix should look like the one in Figure 6.17.

Event	myhealth		pophlth		efficsys		rspnstim		nonmed	
myhealth	Event1-1	☑	Event1-2	☑		☐		☐		
-->	0		0							
pophlth	Event2-1	☑		☐	Event2-3	☑	Event2-4	☑	Event2-5	
-->	0				0		0		0	
efficsys	Event3-1	☑	Event3-2	☑		☐	Event3-4	☑	Event3-5	
-->	0		0				0		0	
rspnstim	Event4-1	☑	Event4-2	☑	Event4-3	☑		☐		
-->	0		0		0					
nonmed	Event5-1	☑		☐	Event5-3	☑		☐		
-->	0				0					
noofdocs	Event6-1	☑	Event6-2	☑	Event6-3	☑	Event6-4	☑		
-->	0		0		0		0			

Figure 6.16: Blank Healthcare Event Matrix.

Event	myhealth		pophlth		efficsys		rspnstim		nonmed	
myhealth	Event1-1	☑	Event1-2	☑		☐		☐		
-->	0		30							
pophlth	Event2-1	☑		☐	Event2-3	☑	Event2-4	☑	Event2-5	
-->	70				10		80		-50	
efficsys	Event3-1	☑	Event3-2	☑		☐	Event3-4	☑	Event3-5	
-->	60		30				90		-50	
rspnstim	Event4-1	☑	Event4-2	☑	Event4-3	☑		☐		
-->	-50		80		30					
nonmed	Event5-1	☑		☐	Event5-3	☑		☐		
-->	-70				-70					
noofdocs	Event6-1	☑	Event6-2	☑	Event6-3	☑	Event6-4	☑		
-->	90		80		30		60			

Figure 6.17: Filled-in Healthcare Event Matrix.

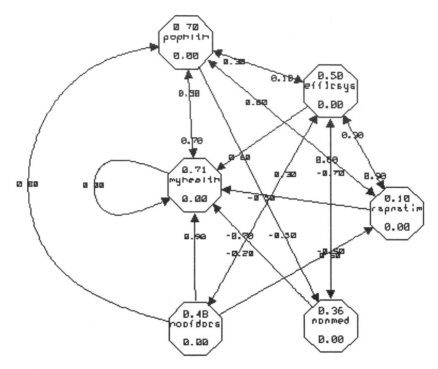

Figure 6.18: Healthcare map displaying event labels.

Drag the matrix to the bottom of the screen so that you can see the map itself. You'll see the weights displayed next to the event arrows (Figure 6.18.)

Before running the map, initialize the history function so you'll have a record of the cycles.

Running the Healthcare Map Cycles

Instead of seeing the state values change in the List box, you can observe them with greater in the Observer.

Select Observe from the Wizard menu and click on the Enable Observer box. The map's calculations may be performed in integer (whole number) arithmetic or in double-precision floating point arithmetic (to 16 decimal places). Double-precision floating point is the default.

To run the cycles until no more changes occur, click on the Start button. Watch the Error window to see the changes become smaller, until the value to 16 decimal places is 0.

Importance of the Healthcare Map

When the cycles stop at 0, the map has reached stability. This has shown one way to understand how the health care system works. Take a look at the final state values as recorded in the historical record. Are they meaningful in terms of this map? Are they meaningful to you?

There are no correct answers to these questions. So Dr. Fuzzy invites you to work with this map by changing state values and event weights, creating or deleting events, and otherwise personalizing the map until it is meaningful for you.

TRAINING A MAP TO PREDICT THE FUTURE

Its ability to be trained shows a similarity between fuzzy cognitive maps and neural networks. What kind of futures do people want to predict? Money making is high on the list. Of course, this has a long and respectable history within computer science. Two of the early computer pioneers, Charles Babbage and his assistant Augusta Ada, Countess of Lovelace, were strongly motivated in their work by the thought of discovering a foolproof system for winning at the race track!

Dr. Fuzzy doesn't guarantee any results for trained the cognitive maps. But some trained maps are being used as tools for stock market traders.

The Scenario

Numerous indexes of stock values for various stock exchanges are published in newspapers every day. Many of these indexes play large roles in the buying and selling behavior of stock traders. Other influential information includes the money supply, the prime lending rate, and the consumer price index.

TABLE 6.9: Names and Descriptions of Stock Market Map.

Name	Description
MoneySup	Money supply
PrimeRte	Prime rate
TodaysAv	Today's 30 stock average
CnsmrPI	Consumer Price Index
TmorwsAv	Tomorrow's 30 stock average

What follows is the good doctor's implementation of a map that uses this type of information to predict tomorrow's value of a theoretical index of 30 industrial stocks.

The States

A states of a trained map are set up the same way as those of a normal map.

Open a new Fuzzy Thought Amplifier™ map, naming it *30stkavg* and equip it with *five* states. Define the states from the information in Table 6.9. Arrange the states in the shape shown in Figure 6.19.

The Events

Create events as shown in Table 6.10.

E-MAIL   A reminder: Predetermined state values train the event
FROM   weights. When the map is run, the weights are fixed and se-
DR. FUZZY lected state values change.

In some situations, the map maker doesn't know what the initial event weights should be. You can select random values, or create some other type of distribution. For the stock market map, Dr. Fuzzy has employed an equal

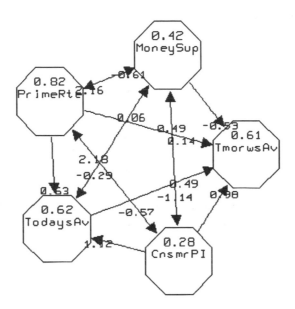

Figure 6.19: States for the stock market cognitive map.

TABLE 6.10: Events for Stock Market Map.

Causing State	Affected State
1	2
1	3
1	4
1	5
2	1
2	3
2	4
2	5
3	1
3	5
4	1
4	2
4	3
4	5

TABLE 6.11: Beginning Weights for Stock Market Map.

Event	Weight
1 → 2	-100
1 → 3	-85
1 → 4	-70
1 → 5	-55
2 → 1	-39
2 → 3	-24
2 → 4	-8
2 → 5	8
3 → 1	24
3 → 5	39
4 → 1	55
4 → 2	70
4 → 3	85
4 → 5	100

distribution of values between -1 and +1 (or -100 and +100). Enter the beginning weights listed in Table 6.11 into the Event Matrix.

Training the Map

Training a fuzzy cognitive map involves compiling sets of historical data that are run through the map one at a time. Each set is cycled until no more changes ("error") occur in the weights or until the last three decimal places oscillate. The Fuzzy Thought Amplifier™ can use a maximum of 25 training sets.

The time interval between training sets determines the time interval of the prediction. For instance, if you use daily training sets, the prediction will also be one day. If you use monthly sets, the prediction interval will also be one month.

To begin the training process, select Training Sets from the Wizard menu. Enter the values in Table 6.12 into the spreadsheet.

When you've completed entering the training sets, click on the Save to File button. The file can then be retrieved for training or for modification, if necessary.

To use the training sets, select Train from the Wizard menu. The Training of Causal Events dialog box will appear on the screen. First, click

TABLE 6.12: Training Sets for the Stock Average Map.

Set No.	States				
	MoneySup	*PrimeRte*	*TodaysAv*	*CnsmrPI*	*TmorwsAv*
1	50	64	33	24	35
2	51	62	35	21	37
3	49	63	37	23	38
4	48	62	38	24	39
5	49	63	39	25	38
6	50	61	38	26	40
7	51	63	40	24	41
8	53	67	41	22	45
9	54	64	45	21	46
10	56	62	46	22	47
11	57	67	47	23	48
12	55	68	48	25	48
13	54	69	48	24	49
14	51	70	49	23	50
15	51	71	50	22	50
16	48	75	50	22	52
17	48	75	52	21	54
18	46	78	54	24	56
19	43	81	56	23	58
20	42	83	58	21	59
21	43	82	59	22	61
22	42	81	61	23	63
23	42	83	63	25	66
24	41	82	66	27	67
25	42	82	67	28	62

on the button that reads Load training set file (lower left corner), which makes the training sets available for the process.

Next, be sure that the window to the right of the Use this file button reads *1*. If not, use the up–down arrows to set it to 1. Click on the Use this training set button.

Click on the Start–Stop button (upper right). This will send the first training set through the map. You can see the error diminish to 0, at which time the cycles will stop. If the three right-hand values oscillate, click the Start–Stop button again to stop the cycles.

Now click on the Use next training set button. The value in the window will advance to 2. Again click on the Start–Stop button until the error diminishes.

Continue the process until all 25 sets have been used to train the map. Then click on OK.

Now you can use your trained map. But first, take a look at the Event Matrix. You'll see that the weights have changed considerably from the initial values.

Predicting the Future

The key to predicting the future with this map is your next operations in the List box, which you should display. Your final training set was from "yesterday."

First, fix the values for the MoneySup, PrimeRate, and CnsmrPI by clicking on their Fix boxes. Red check marks will appear.

Now change the TodaysAv value from "yesterday's" 67 to "today's" 62. Fix that value.

Now click on the Step Forward icon. The cycle will change the TomrwsAv value to 61. That means that the fuzzy cognitive map predicts that the final value tomorrow of the 30-stock average will be 61.

What can you do with the information? Dr. Fuzzy says, "It's up to you!"

Now that you've seen this hypothetical map in action, feel free to use it with real information from the financial pages. But Dr. Fuzzy doesn't guarantee anything!

HOW THE FUZZY THOUGHT AMPLIFIER™ WORKS

The Fuzzy Thought Amplifier™ provides two computational methods, Definition and Incremental.

Definition Method

The Definition method has been described by Bart Kosko and others. Each state's value is completely defined anew during each forward step (cycle) as long as the map is dynamic. Each state's value is the result of taking all the causal event weights pointing into the state, multiplying each weight by the weight of the event's causing state, and adding up all the results of these multiplications. The results are then squashed so that the result is between 0 and 1 (or 0 and 100%).

This multiply-and-sum process is a linear operation. The result lies between the range -StateCount to +StateCount, where StateCount is the number of states in the map.

The squashing operation itself is nonlinear, calculated from a mathematically defined "logistic function." This function symmetrically around 0 on the input side and around .5 on the output converts the input of -State-Count to +StateCount to the output of 0 to 1. The equation is

$$\text{Activation} = \frac{1}{1 + e^{-(\text{Summation} \times \text{Gain})}}$$

Incremental Method

The incremental method draws on cellular automata techniques. Each state value is a modification from the previous value during each forward step of the dynamic map. Each state's value is the result of taking all the event weights that affect the state, multiplying each by the causing state's value, summing all the results of these multiplications, incrementing the previous presquashed value by a fraction of this summation result, and then squashing the results of the incrementation so that it is between 0 and 1 (0 and 100%).

Training Function

The training function works by taking each incoming event to a state from a Causing State and multiplying it by the Causal Weight. This produces the Vector Sum.

The State Sum is the sum of ALL of the Causing States that direct events to one state, which is what the state value would be. The difference between that and what the value really is an Error. The cycling continues until the error is gradually corrected.

The value of each Causing State divided by the State Sum is the normalizing process. Multiply the normalized value by the error, which is they added to the incoming weight.

This process is repeated for each state until the error goes to 0.

CONCLUDING THOUGHTS

You have now had in-depth adventures with the three principal fuzzy architectures.

The fuzzy expert system has a secure and growing place in the commercial world for control and other applications. The Fuzzy Knowledge Builder™ is a tool you can use to put a fuzzy expert system to work for your own purposes.

Decision making is an intuitive process we've all used since infancy for matters great and small. The Fuzzy Decision Maker™ is a tool for organizing your personal or work-related decision process. This is the first time a fuzzy decision maker has been presented to a general audience, and we (and Dr. Fuzzy) hope you will find innovative ways to use it.

The fuzzy cognitive map is a tool in search of practical use. With the Fuzzy Thought Amplifier™, we present the first implementation for a general audience and invite you to experiment with it.

To round out the picture, Appendix E describes the two other fuzzy architectures that we know of. FLOPS is a fuzzy extension of a rule-based environment called OPS5, which is well known in artificial intelligence circles.

Fuzzy state machines are attracting increased research interest for time-dependent problem solving. Dr. Fuzzy hasn't implemented this architecture—yet.

E-MAIL FROM DR. FUZZY   I hope you'll generalize your view of world from the crisp to the far more realistic fuzzy mode and continue with this adventure.
Best wishes!

APPENDIX A

FUZZY ASSOCIATIVE MEMORY (FAM)

Most fuzzy systems (see Chapter 3) represent inputs and outputs as membership functions whose interactions are the bases for rules and a fuzzy action surface. Inference involves the firing of individual rules.

There's another way to create an action surface for multiple inputs *and* multiple outputs. Called the *compositional method*, it was actually the original method. The fuzzy input and desired output ranges are based on fuzzy set values and used to create a matrix called a *fuzzy associative memory* (FAM). When actual input values enter the system, the associative memory becomes the fuzzy action surface. The entire memory fires at once, producing multiple outputs.

Prepared for all situations, Dr. Fuzzy has created a calculator for the compositional method, FAMCalc. To open FAMCalc, click on the Convertible icon.

E-MAIL
FROM
D R. FUZZY

When would you want a multiple-input system that produces multiple outputs? Here's an example.

Suppose you have a graphics enhancement problem
involving the gray scale. In this case, you could have four
one-pixel inputs whose gray scale values you want to automatically change. By putting them through the FAM, the
output is a revised gray hue for each of the four pixels.

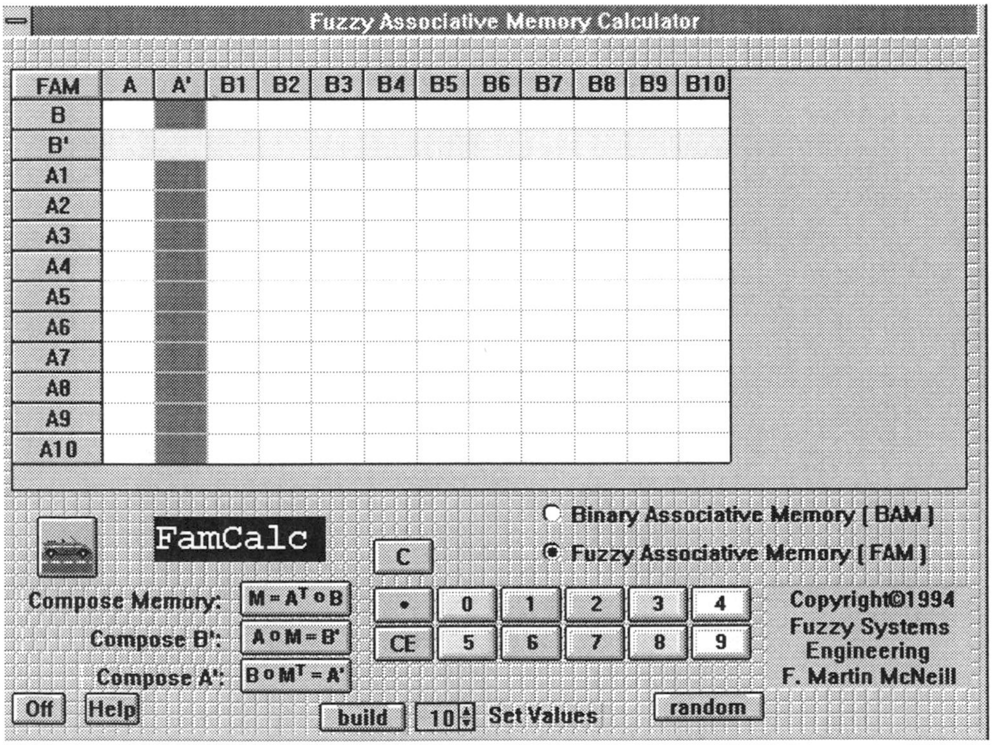

Figure A.1: FAMCalc opening screen.

FAMCALC

The heart of FAMCalc (Figure A.1) is the blank matrix with room for as many as 10 inputs in the left-hand column (A1 through A10) and 10 outputs on the top row (B1 through B10). Each input and output represents a fuzzy set and can have any value between 0 and 1.

Three operator keys are on the lower left side of the calculator. The Compose Memory key constructs the FAM after you have entered the input and output values. B′ is the output of passing A through the FAM, which is initiated with the Compose B′ button. The B′ values appear in the yellow row. The final function is Compose A′, which is the result of passing B backwards through the FAM, a feedback mechanism that may be useful in some situations. A′ is displayed in the magenta column.

In the lower right-hand corner is the calculator keypad with numbers 0–9, Clear and Clear Entry buttons, and an Example button. The Fuzzy Associative Memory (FAM) button is for fuzzy calculations and Binary Associative Memory (BAM) is for crisp ones.

Use the button with up- and down-arrow keys to choose the number of inputs and outputs (they must be the same), then click on the Build button to configure the matrix.

COMPOSING A MEMORY

To see how FAMCalc works, create a four-by-four matrix with the arrow keys and Build. Once the matrix is reconstructed, enter these input (A) and output (B) values:

A	A′	B1	B2	B3	B4
B		.2	.4	.7	.9
B′					
A1					.1
A2					.3
A3					.6
A4					.8

FAM	A	A'	B1	B2	B3	B4
B'			0.20	0.40	0.70	0.90
B						
A1	0.10		0.10	0.10	0.10	0.10
A2	0.30		0.20	0.30	0.30	0.30
A3	0.60		0.20	0.40	0.60	0.60
A4	0.80		0.20	0.40	0.70	0.80

Figure A.2: Composing the FAMCALC matrix.

Now click on the Compose Memory key, producing the matrix shown in Figure A.2.

To use the FAM, click the Compose B' key. The results for the actual output values are very close to the design output:

B	.2	.4	.7	.9
B'	.2	.4	.7	.8

When you feed back the B values through the FAM by clicking on the Compose A' key, you'll find that A' is identical with A.

To see how crisp values fare in an associative memory, click on the Binary Associative Memory button and watch the input values change to 0s and 1s. The same thing will happen to the memory itself and to B' and A' when you click on their operator keys. The results are shown in Figure A.3.

E-MAIL FROM DR. FUZZY
You can change the values from fuzzy to crisp by clicking on the BAM button. But you can't change the crisp values back to fuzzy (because there are too many fuzzy options for each crisp 0 or 1).

If you want to experiment with some preset fuzzy inputs and outputs, click on the Example button. You can use it with any size matrix.

Try this. After composing the memory, change several input values and click on Compose B'. Notice the difference between B' and the design outputs in B. Now click on Compose A' and see more differences.

CREATING A MEMORY

There's a second way to use FAMCalc—by entering the associative memory values, then entering the inputs and having the FAM produce the outputs. For instance, enter the matrix values in the first example, then enter the inputs.

Just click on each blank FAM cell, then enter the values from the calculator keypad or by hand. The results should look like those in Figure A.4. Now press Compose B'. As you can see, the results are the same as in the original example.

When you're finished with FAMCalc, click on the OFF button in the lower left-hand corner.

FAM	A	A'	B1	B2	B3	B4
B'			0.00	0.00	1.00	1.00
B			0.00	0.00	1.00	1.00
A1	0.00	0.00	0.00	0.00	0.00	0.00
A2	0.00	0.00	0.00	0.00	0.00	0.00
A3	1.00	1.00	0.00	0.00	1.00	1.00
A4	1.00	1.00	0.00	0.00	1.00	1.00

Figure A.3: FAMCalc in the crisp mode.

FAM	A	A'	B1	B2	B3	B4
B'						
B						
A1			0.10	0.10	0.10	0.10
A2			0.20	0.30	0.30	0.30
A3			0.20	0.40	0.60	0.60
A4			0.20	0.40	0.70	0.80

Figure A.4: A hand-entered matrix.

HOW FAMCalc WORKS

How does FAMCalc do its job? Dr. Fuzzy reveals the two-step method:

Step 1

Look again at the values in Figure A.4. Compare the value of A1 with each memory value in the same row:

$$A1 \quad .1 \quad | \quad .1 \quad .1 \quad .1 \quad .1$$

Take the *minimum* in each pairing in a *row*, giving you these *pairwise minima*:

$$.1 \quad .1 \quad .1 \quad .1$$

That one was almost too easy. What would be the pairwise minima for the next row,

$$A2 \quad .3 \quad | \quad .2 \quad .3 \quad .3 \quad .3$$

Again, it's an easy answer:

.2   .3   .3   .3

Now find the pairwise minima for the last two rows,

A3	.6	|	.2	.6	.6	.6
A4	.8	|	.2	.4	.7	.8

The results are

.2	.6	.6	.6
.2	.4.	.7	.8

So all the pairwise minima are:

.1	.1	.1	.1
.2	.3	.3	.3
.2	.6	.6	.6
.2	.4.	.7	.8

Step 2

Take the *maximum* of each *column* in the pairwise minima, producing

.2   .6   .7   .8

These maxima are B', the outputs.

APPENDIX B

FUZZY SETS AS HYPERCUBE POINTS

SETS AS POINTS

You've already seen how fuzzy sets can be represented by membership functions. There's another way to consider them—as points in a theoretical structure called a *hypercube*. This idea was invented and proven by fuzzy logic theoretician Bart Kosko.

A hypercube can have any number of dimensions, but for practical purposes here, Dr. Fuzzy recommends thinking about the 2-D hypercube—a square.

Take a 2-D hypercube—a square (a "hypocube"?)—and place a point at each node or vertex and one in the center (Figure B.1). You can think of the square as a set of all fuzzy sets—a superset. The square is like a graph with an x axis and a y axis. This means you can locate any point in it with two coordinates. So a point in a 2-D hypercube is the location of a set represented by a two-valued membership function.

E-MAIL FROM DR. FUZZY

A hypercube is a geometric shape of *n* dimensions with identical edges. For example, a two-dimensional hypercube is a square, whose arms of equal length. The places where the arms meet are called *nodes* or *vertexes*.

A three-dimensional hypercube (a cube) adds an edge the length of the square's arm into the third dimension from each vertex.

The next dimension can only be seen in the mind's eye. Try to visualize the 4-D hypercube as a 3-D cube with another cube added at each vertex.

The purpose of talking about hypercubes is to let you think of the maximum generalization of fuzziness.

In fuzzy logic, *a hypercube has as many dimensions as the set has values.* A set with 2 values has a 2-D hypercube. A 20-value set has a 20-D hypercube. An *n*-valued set has an *n*-D hypercube.

The vertex points represent *crispness*—coordinates must be either 0 or 1. Each side of the square represents a fuzzy range between 0 and 1. You can also move along diagonals from crisp 0 or 1 through the square to another crisp 0 or 1.

What does this mean? The square is a device for visualizing the relationship of the fuzzy set A to its complement A^c, as defined in Chapter 2 for either their intersection $(A \cup A^c)$, or their union $(A \cap A^c)$. The center point represents a set with *maximum fuzziness*—the place where, as you saw in Chapter 2, $A = A^c$.

Dr. Fuzzy was at a carnival recently and noticed how much one of the gameboards on the midway resembled a fuzzy 2-D hypercube. The players have to slide a disk so that it stops somewhere on the board. If the disk lands on one of the corners, the lucky player wins a big prize. If it lands in the center, the prize is kind of puny. If it lands somewhere in between, the size of the prize depends on how close the disk is to one of the vertexes—the closer the disk is, the bigger the prize. The Good Doctor immediately dubbed the game Fuzzy Shuffleboard.

How to determine the set's fuzziness? The games master needed help. Always willing to assist an entrepreneur, Dr. Fuzzy immediately hot-footed it home and invented a calculator to solve sets-as-points problems. The doctor named it KoskoCalc, in honor of Bart Kosko. Open KoskoCalc by clicking on the racing car icon.

USING KOSKOCALC

KoskoCalc (Figure B.2) has a blank matrix with room for as many as 25 values for each of two sets, A and B. In this case, the hypercube will be 25-D. Ten values are displayed on the opening screen. If you want more (or fewer) than 10, use the Build system at the center bottom. Click on the up or down arrows until the desired number of values appears. Then click on Build to remake the matrix. If you Build more than 10 values, a horizontal scroll bar below the matrix lets you display all of them.

You can enter your own values, or click on the Random button on the lower right corner for Dr. Fuzzy's selections. The operator keys are on the lower left side of the calculator.

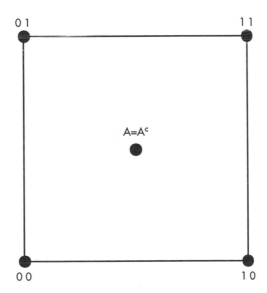

Figure B.1: A 2-D hypercube.

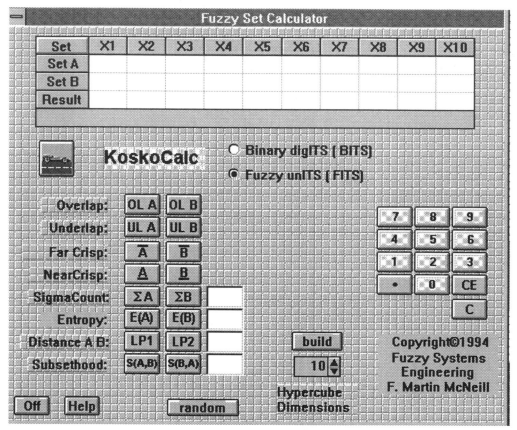

Figure B.2: KoskoCalc opening screen.

INTERACTION OF A SET AND ITS COMPLEMENT

Now you can mark the location of a different fuzzy set in the hypercube (Figure B.3). Its degree-of-fuzziness values are .3 and .8.

You can enter this set's values in KoskoCalc. To begin, Build a matrix with two values by clicking on the up–down arrows until the number 2 appears. Then click on Build.

Next, enter these values (from Figure B.4) in the matrix:

	X1	X2
Set A	.3	.8

(We'll get to Set B later.)

The hypercube's vertexes have what's called *minimum entropy* and its center has *maximum entropy*. The term *entropy* comes from the field of thermodynamics, where it means the amount of disorder in a system.

E-MAIL FROM DR. FUZZY

It's also used in information science to mean the amount of information in a message. The greatest uncertainty (meaning the least information) means the maximum entropy. The least uncertainty and most information is the minimum entropy.

The "information" meaning of entropy is also used in fuzzy logic. Kosko says that fuzzy entropy is the measure of fuzziness.

The entropy of a set named Set A can be calculated by clicking on the $E(A)$ button on KoskoCalc. The $E(B)$ button calculates the entropy for another set named Set B. The result is displayed in the box next to the button.

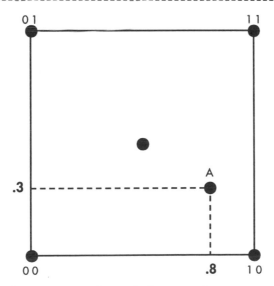

Figure B.3: One fuzzy set in the 2-D hypercube.

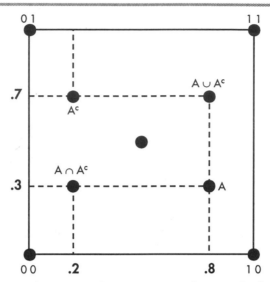

Figure B.4: Set A and its complement, together with their overlap and underlap.

Once you've identified a set's degree of fuzziness (as in Figure B.4), you can determine the complement A^c, their intersection ($A \cap A^c$), which is also called *overlap*, and their union ($A \cup A^c$), called *underlap*. You can also calculate the set's distance from the nearest and farthest crisp vertexes.

The complement of Set A is

$$A^c = (.7 \ .2)$$

To determine Set A's overlap ($A \cap A^c$), click on the OLA button. (When you have a Set B, calculate its overlap ($B \cap B^c$) by clicking on the OLB button.) The answer appears in the Results column of the matrix:

$$A \cap A^c = (.3 \ .2)$$

To calculate the underlap of Set A ($A \cup A^c$), click on the ULA button. The answer appears in the Results column of the matrix. To calculate the underlap of a Set B ($B \cup B^c$), click on the ULB button. The answer is

$$A \cup A^c = (.7 \ .8)$$

These operations are plotted on Figure B.4.

FAR CRISP AND NEAR CRISP

To determine the precise location of a set, you can calculate its distance from the farthest crisp vertex and from the nearest one. These calculations look easy in 2-D, but if you're working with a 20-D hypercube, for example, you need major help.

To determine Set A's farthest crisp vertex, click on the $\overline{A}$ button. For the nearest crisp vertex, click on $\underline{A}$. The answer appears in the Results column of the matrix. (If you have a Set B, you can calculate the same values for it by clicking on the $\overline{B}$ button and then the $\underline{B}$.)

How are the calculations performed? Think of each value individually. For instance, Set A's X1 value of .3 is located vertically between the crisp 0 and crisp 1 vertexes (Figure B.5). X1's farthest crisp vertex is 1 and its nearest crisp vertex is 0.

The X2 value of .8 is located horizontally between crisp 0 and crisp 1 vertexes (Figure B.6).

So X2's farthest crisp vertex is 0 and its nearest crisp vertex is 1.

MEASURING A SET'S SIZE

The size (cardinality) of the fuzzy set as point is its distance from its vertex of origin, 0 0 (see Figure B.7). It has several other names. It's called M or the

---

$$\text{fuzzy entropy} = \frac{\text{overlap}}{\text{underlap}}$$

**E-MAIL
FROM
DR. FUZZY**

The formal equation for the entropy of Set A is

$$E(A) = \frac{M(A \cap A^c)}{M(A \cup A^c)}$$

M stands for measure.

Figure B.5: Position of Set A's X1 value.

Figure B.6: Position of Set A's X2 value.

measure of A, $M(A)$. It's also known as the sigma (Σ) count or the Hamming norm (ℓ^1).

This Hamming or "city block" method is the preferred way to measure, rather than the alternative, which would be the direct diagonal from the originating vertex (0 0) to Set A's point.

To calculate Set A's sigma count, click on the ΣA button. The result is shown in the box next to the button. (For Set B's sigma count, click on ΣB.)

E-MAIL
FROM   Here's another way to think of it.
D R. FUZZY

INTERACTION OF TWO FUZZY SETS

You can also perform operations on several sets within a hypercube. Kosko-Calc provides for two sets. You can enter the following values for Set B:

	X1	X2
Set B	.9	.4

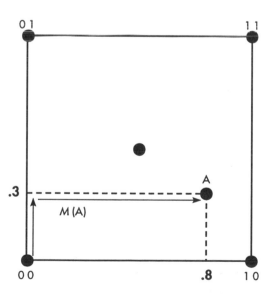

Figure B.7: The measure of Set A or its sigma count.

Among the operations you can perform on them is to measure their distance from each other (useful if two people are playing Fuzzy Shuffleboard) and the extent to which each set is a subset of the other. Set A and Set B are graphed in Figure B.8.

In other words, you can stick to the sidewalks or cut catty-corner across the grass.

In math terms, the sidewalks around the block measurement is called ℓ^1, which is actually the

$$X1 + X2 + \dots$$

If you had n values and an n-D hypercube, it would be

$$X1 + X2 + \dots + Xn$$

or more formally

E-MAIL FROM DR. FUZZY

$$\sqrt[1]{X1^1 + X2^1 + Xn^1}$$

The catty-corner method is called ℓ^2. For an n-D hypercube, the math representation is

$$\sqrt[2]{X1^2 + X2^2 + Xn^2}$$

If this strikes a high school geometry chord, it may be because this is really the Pythagorean theorem. The diagonal measurement is like the hypoteneuse of a right triangle, and the relationship is considered a generalization of the (crisp) Pythagorean theory.

Other ways to measure are possible. For instance, the ℓ^3 method is the cube root of the sum of the cubes of X1, X2, ...

Distance

The distance between Set A and Set B can be measured in much the same way as a set's sigma count. The sets' "city block" distance (Figure B.9) is called the *Hamming distance* or ℓ^{P1} (the equivalent of ℓ^1).

    To perform this operation, click on KoskoCalc's LP1 button. The result appears in the box next to the button.

    You can also measure the between-sets distance by the "catty corner" method, or ℓ^{P2} (the equivalent of ℓ^2) (Figure B.10). Click on the calculator's LP2 button to perform this measurement.

Subsethood

Each set may be a subset of the other. How? If there's a region where the two sets intersect: A B. The point of intersection is subset B* of Set B and subset A* of Set A (Figure B.11).

    In general, the degree to which Set A is a subset of Set B is

$$\frac{\text{the size of the intersection of Set A and Set B}}{\text{the size of Set A}}$$

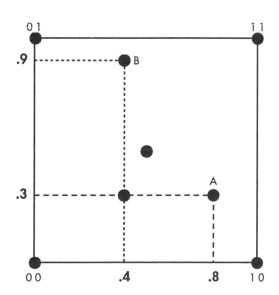

Figure B.8: Sets A and B.

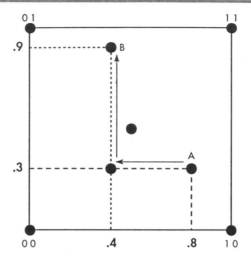

Figure B.9: Measuring the fuzzy sets' distance by ℓ^{P1} ("city block").

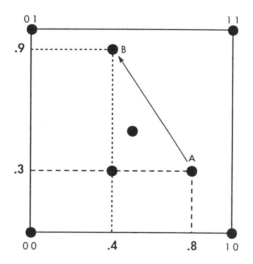

Figure B.10: Measuring the fuzzy sets' distance by ℓ^{P2} (diagonal method).

The formal formula is

$$S(A, B) = \frac{M(A \cap B)}{M(A)}$$

Kosko has demonstrated the similarity between this formula and one for Bayesian probability theory. First he shows that

$$M(A \cap B)$$

implies

$$M(B)S(B,A)$$

E-MAIL FROM DR. FUZZY He then shows that the full fuzzy equation

$$S(A, B) = \frac{M(A \cap B)}{M(A)}$$

implies the Bayesian

$$S(A, B) = \frac{M(B)S(B, A)}{M(A)}$$

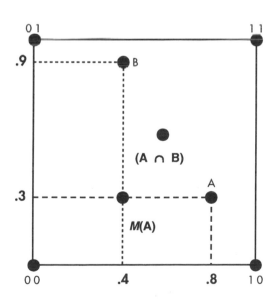

Figure B.11: Subsethood, Sets A and B.

The degree to which Set B is a subset of Set A is

$$\frac{\text{the size of the intersection of Set A and Set B}}{\text{the size of Set B}}$$

or

$$S(A, B) = \frac{M(A \cap B)}{M(B)}$$

Calculate the subsethood of (A,B) and (B,A) by clicking on the appropriate buttons S(A,B) or S(B,A). The result appears in the box next to the button.

Figure B.10 shows the results of calculations for Set A's overlap, sigma count, and entropy, the LP1 calculation for Distance AB, and the subsethood S(A,B).

When you're finished with KoskoCalc, click on the OFF button to return to the main calculator menu.

APPENDIX C

DISK FILES AND DESCRIPTIONS

This appendix contains an annotated list of the files on the accompanying disk.

LIBRARY FILES

Five run-time libraries for *.exe* files are included. They are installed automatically in the Windows\system directory.

- tbpro1w.dll
- tbpro2w.dll
- tbpro3w.dll
- tbpro5w.dll
- ctl3dv2.dll

DR. FUZZY'S CALCULATORS

All of the calculators are in a single executable file:

fuzzcalc.exe

The on-line help system is in the file fcalchlp.hlp.

FUZZY KNOWLEDGE BUILDER™ FILES

The book version of the Fuzzy Knowledge Builder™ application, described in Chapter 4, is in the executable file

fuzzykb.exe

The application's run-time library is

buildint.exe

The Fuzzy Knowledge Builder™ on line help system is contained in the file

fkbhlp.hlp

Example Knowledge Base

An example knowledge base in the application's data script for the BASIC inference engines is contained in the file

test.fdt

Example Inference Engines

Examples of inference engines in several languages are provided.

Motorola 68HC05 Assembly Language

A simple inference engine in Motorola 68HC05 assembly language is in the file

ie05.asm

Basic

Two inference engines written in Basic are provided. They run in DOS QuickBasic. The Simple engine is

fuzzy1.bas

The Fast engine is

fuzzy2.bas

Test files for use with the Basic inference engines are in the files

- test.fam
- test.rul

C Language

The file

cie.c

contains C language code fragments of a C inference engine and an example of the use of the ieTag structure.
  The file

testie.fic

contains an example of cie.c code fragments.
  The file

trktrl.c

is an example inference engine with an embedded knowledge base from Fuzz-C (Bytecraft, Inc.).
The file

trktrl.fuz

is a Fuzz-C example from the Fuzzy Knowledge Builder™.

Example Problems

Several examples that supplement those in Chapter 4 are in disk files. Those with the extension *.fam* contain the structure and those ending in *.rul* contain the rules.

Multiplication

A simple example of using the Fuzzy Knowledge Builder™ is that of multiplication. The files are

- multiply.fam
- multiply.rul

Truck Backing Up

A classic fuzzy expert system problem, first devised and solved by Bart Kosko, is how to back up and park a large tractor-trailer. The Fuzzy Knowledge Builder™ implementation is contained in the files

- truck.fam
- truck.rul

The file

trktrlt.f05

contains a related example knowledge base in Motorola 68HC05 assembly language for the ie05.asm inference engine.
The file

trktrlt.fdt

contains a related input data script knowledge base.

Random Example

A random example is contained in the files

- random.fam
- random.rul

FUZZY DECISION MAKER™

The executable file for the book version of the Fuzzy Decision Maker™ is

fuzzydm.exe

On-line help is in the file

fdmhlp.hlp

Several example problems solved with the Fuzzy Decision Maker™ are provided on the disk.

Choosing a College

The college selection scenario, discussed in Chapter 5, is in the file college.dec

In addition, the following examples are provided in files.

Legal Problem

An example of solving a legal problem is in the file

legalde.dec

Unemployment

Making a decision about finding a new job is explored in

employ.dec

Financial Planning

An aspect of financial planning is contained in the file

finplan.dec

Changing Residence

The criteria that might go into the decision to change your residence is explored in

move.dec

FUZZY THOUGHT AMPLIFIER™

The executable file for the book version of the Fuzzy Thought Amplifier™ (Chapter 6) is in the file

fuzzyta.exe

The on-line help system is in

fcmhlp.hlp

Two cognitive maps discussed in Chapter 6 are provided on the disk. The three simple maps showing end behaviors are in the file

example.fcm

The Cat-and-Plant environmental system is in the file

catplant.fcm

This same map, but in color and dynamic state sizes, is in

catplnt2.fcm

README FILE

The list of disk files is in

readme.txt

APPENDIX D

INFERENCE ENGINE PROGRAMS

This appendix contains examples of inference engines written in QBASIC, C language, Fuzz-C (Bytecraft, Inc.), and the Motorola 68HC05 assembly language. The QBASIC engines (written by Martin McNeill) are those used in Chapter 4 and can be further investigated on any MS-DOS machine, since QBASIC is a part of all versions of DOS 5 and DOS 6.

QUICKBASIC SIMPLE INFERENCE ENGINE

This inference engine is *fuzzy1.bas*, and can be tested with the file *test.fdt* that's on the disk, or with any file you name *test.fdt*. You can also substitute any other file name with the .fdt extension. In that case, you should change the file references in fuzzy1.bas. These are in bold type in the program listing.

Fuzzy1.bas can handle triangular and trapezoidal membership functions. It's called simple because it's a "no frills" program and tests all the rules, active or not.

The code for fuzzy1.bas follows. (To accommodate the printed page, long lines that must carry over are broken with a double slash, //.)

```
'Fuzzy Exersizer
'Reads TEST.FDT
'Allows
DECLARE SUB ShowOutput ()
DECLARE SUB ReadKnowledgeBase ()
DECLARE SUB FuzzyMap ()
DECLARE SUB GetInput ()
DECLARE SUB DisplayOutput ()
COMMON Rule() AS INTEGER
'Global. From XXXX.FDT.
'The rule array size is defined at knowledge base read time.
DIM SHARED NumInputs AS INTEGER
'Global. From XXXX.FDT.
'The Number of Inputs or input dimensions(1 to 5)
DIM SHARED NumOutputs AS INTEGER
'The Number of Outputs (1 to 2) Each output implies a separate estimation
'surface. The input permutations are the same.
DIM SHARED Credibility AS INTEGER
'Ignore Credibility now. The inference engine does not use it.
DIM SHARED NumRules AS INTEGER
'Global. From XXXX.FDT.
'The number of rules to be read in. DIM SHARED NumFS(6) AS INTEGER
'Global. From XXXX.FDT.
'The number of fuzzy sets on each input (0 to 4) and each output (5 to 6)
'This may vary from 1 to 11 on each input or output.
'Used to control readin and inference engine.
DIM SHARED PWFS(6, 10, 3, 1) AS INTEGER
'Global. From XXXX.FDT.
'The piecewise fuzzy sets are stored here:
'0 to 6: The input and output indicies, same as NumFS(6)
'0 to 10: The fuzzy set index on each input or output dimension.
'    The maximum value is the same as stored in NumFS(6).
'0 to 3: The number of points on each fuzzy set in each dimension.
'    The values may be: 1 = Singleton, 3 = Triangle, 4 = trapezoid.
'    Same as value stored in SHAPE(6).
'0 to 1: The number of values needed to define each vertex point of the
'    fuzzy sets. 1 = Singleton, 2 = Other.
DIM SHARED FInValue(4) AS INTEGER
'Global.
'Values into the inference engine. From GetInput sub or from
'    your application.
DIM SHARED FOutValue(1) AS INTEGER
```

```
'Global.
'Values from the inference engine. To ShowOutput sub or to
'   your application.
DIM SHARED SHAPE(6) AS INTEGER
'Global. From XXXX.FDT.
'   The values may be: 1 = Singleton, 3 = Triangle, 4 = trapezoid.
DEFINT A-Z

DO
    CLS
    PRINT "Fuzzy Work": PRINT
    COLOR 15, 0: PRINT "  R"; :  COLOR 7, 0: PRINT "ead Knowledge Base"
    COLOR 15, 0: PRINT "  F"; :  COLOR 7, 0: PRINT "uzzy Map"
    COLOR 15, 0: PRINT "  I"; :  COLOR 7, 0: PRINT "nput Values"
    COLOR 15, 0: PRINT "  S"; :  COLOR 7, 0: PRINT "how Output Values"
    COLOR 15, 0: PRINT "  Q"; :  COLOR 7, 0: PRINT "uit"
    PRINT : PRINT "Select: "

    ' Get valid key
    DO
            Q$ = UCASE$(INPUT$(1))
    LOOP WHILE INSTR("BRFISQ", Q$) = 0

' Take action based on key
    CLS
    SELECT CASE Q$
        CASE IS = "R"
             PRINT "Reading . . ."
            OPEN "TEST.FDT" FOR INPUT AS #1
            DO UNTIL EOF(1)
                INPUT #1, FileString$
                SELECT CASE FileString$
                CASE "NUM_INPUTS"
                    INPUT #1, Count$
                    NumInputs = VAL(Count$)
                CASE "CREDIBILITY"
                    INPUT #1, Count$
                    Credibility = VAL(Count$)
                CASE "INPUT1"
                    INPUT #1, Count$
                    NumFS(0) = VAL(Count$)
                CASE "INPUT2"
                    INPUT #1, Count$
                    NumFS(1) = VAL(Count$)
                CASE "INPUT3"
                    INPUT #1, Count$
                    NumFS(2) = VAL(Count$)
                CASE "INPUT4"
                    INPUT #1, Count$
                    NumFS(3) = VAL(Count$)
```

```
                    CASE "INPUT5"
                            INPUT #1, Count$
                            NumFS(4) = VAL(Count$)
                    CASE "NUM_OUTPUTS"
                            INPUT #1, Count$
                            NumOutputs = VAL(Count$)
                    CASE "OUTPUT1"
                            INPUT #1, Count$
                            NumFS(5) = VAL(Count$)
                    CASE "OUTPUT2"
                            INPUT #1, Count$
                            NumFS(6) = VAL(Count$)
                    CASE "NUM_RULES"
                            INPUT #1, Count$
                            NumRules = VAL(Count$)
                    CASE ELSE

                    END SELECT
            LOOP
            CLOSE #1
            REDIM SHARED Rule(NumFS(0), NumFS(1), NumFS(2), NumFS(3),//
NumFS(4)) AS INTEGER
            ReadKnowledgeBase               PRINT "Fuzzy Mapping . . ."
        CASE IS = "F"
            PRINT "Fuzzy Mapping . . ."
            FuzzyMap
            ShowOutput
        CASE IS = "I"
            PRINT "Inputing . . ."
            GetInput
        CASE IS = "S"
            PRINT "Outputing . . ."
            ShowOutput
        CASE ELSE
    END SELECT
LOOP UNTIL Q$ = "Q"
END

'Display the inferred outputs previously calculated.
SUB DisplayOutput
    SCREEN 0, 0      ' Set text screen.
    DO           ' Input titles.
        CLS
        LOCATE 7, 1
        PRINT "Output 1: "; FOutValue(0)
        IF NumInputs > 1 THEN
            PRINT "Output 2: "; FOutValue(1)
        END IF
        PRINT "OK (Y to continue, N to change)? ";
        LOCATE , , 1
```

```
      OK$ = UCASE$(INPUT$(1))
   LOOP UNTIL OK$ = "Y"

END SUB
```

```
'Inference Engine.
'Up to 5 inputs and 1 output supported
'The crisp input value on each input dimension is tested for nonzero value.
'If a nonzero value occurs, it's value and which fuzzy set it is in is
'registered.
'Then all the rules with all nonzero fuzzy sets are "fired" to provide
'the associated output fuzzy set and the minimum input hit value is used to
'truncate the output fuzzy set. The truncated output fuzzy sets areas are
'summed and the center of gravity of each of the primative pieces times
'the primative area(first moment) is summed.
'Primative areas are triangles and rectangles.
'Then the area sumation results is divided into the first moment summation
'to give the overall center of gravity. This is the crisp output of
'inference process.
SUB FuzzyMap
DIM GrndArea AS LONG, GrndFirstMom AS LONG
'Local.
'Accumulation variables for output pieces.
DIM FuzSum AS LONG
'Local.
'The minimum nonzero membership value of a rule that has fired.
DIM MFHIT(0 TO 4, 0 TO 10) AS LONG
'Local.
'The membership values of the crisp inputs in each fuzzy set in each input
'dimension. Most are zero.
DIM Consequent AS LONG
'Local.
'The rule consequent for the particular rule that is firing.
'Firing means all fuzzified inputs are nonzero on this rule.
DIM Area1 AS LONG, FirstMoment1 AS LONG
'Local.
'Left hand triangle shape area of output truncated trapezoid.
'This shape area times the Center of Gravity (First Moment) of this shape.
DIM Area2 AS LONG, FirstMoment2 AS LONG
'Local.
'Center rectangle shape area of output truncated trapezoid.
'This shape area times the Center of Gravity (First Moment) of this shape.
DIM Area4 AS LONG, FirstMoment4 AS LONG
'Local.
'Right hand triangle shape area of output truncated trapezoid.
'This shape times the Center of Gravity (First Moment) of this shape.
DIM Area AS LONG, FirstMom AS LONG
'Local.
'Intermediate sumation variables. May not be needed.
DIM Inter1 AS LONG, Inter2 AS LONG
```

```
'Local.
'Used in the defuzzification process. Stores the two output dimension values
'  of the trapezoidal peak after truncation. (Triangles turn into trapezoids
'  after truncation.
DIM HitsonDim(4) AS LONG
'Local.
'Used to register the number of nonzero input hits in each input dimension.
DIM FSHit(4, 10) AS LONG
'Local.
'Used to register the input dimension and fuzzy set in which the fuzzified
'input is nonzero.


' COMPUTE Hit Values on memberships */

FOR x = 0 TO NumInputs - 1
    IF (SHAPE(x) = 3) THEN    'Trangles
        FOR y = 0 TO NumFS(x) - 1
            IF (FInValue(x) <= PWFS(x, y, 0, 0) OR FInValue(x) //
>= PWFS(x, y, 2, 0)) THEN
                MFHIT(x, y) = 0  'Missed fuzzy set entirely
            ELSEIF (FInValue(x) //
<= PWFS(x, y, 1, 0) AND PWFS(x, y, 0, 1) = 255) THEN
                MFHIT(x, y) = 255  'Hit high left end
            ELSEIF (FInValue(x) >= PWFS(x, y, 1, 0) AND //
PWFS(x, y, 2, 1) = 255) THEN
                MFHIT(x, y) = 255   'Hit high right end
            ELSEIF (FInValue(x) <= PWFS(x, y, 1, 0)) THEN
                MFHIT(x, y) = ((FInValue(x) - PWFS(x, y, 0, 0)) //
* 255&) / (PWFS(x, y, 1, 0) - PWFS(x, y, 0, 0))
                'Hit is on left slope and between 0 and 255
            ELSE
                MFHIT(x, y) = ((PWFS(x, y, 2, 0) - //
FInValue(x)) * 255&) / (PWFS(x, y, 2, 0) - PWFS(x, y, 1, 0))
                'Hit is on right slope and between 0 and 255
            END IF
        NEXT y
    ELSEIF (SHAPE(x) = 4) THEN 'Trapezoid
        FOR y = 0 TO NumFS(x) - 1
            IF (FInValue(x) <= PWFS(x, y, 0, 0) OR FInValue(x) //
>= PWFS(x, y, 3, 0)) THEN
                MFHIT(x, y) = 0
            ELSEIF (FInValue(x) //
<= PWFS(x, y, 1, 0) AND PWFS(x, y, 0, 1) = 255) THEN
                MFHIT(x, y) = 255
            ELSEIF (FInValue(x) >= PWFS(x, y, 1, 0) AND //
PWFS(x, y, 3, 1) = 255) THEN
                MFHIT(x, y) = 255
            ELSEIF (FInValue(x) <= PWFS(x, y, 1, 0)) THEN
                MFHIT(x, y) = ((FInValue(x) - PWFS(x, y, 0, 0)) * //
255&) / (PWFS(x, y, 1, 0) - PWFS(x, y, 0, 0))
```

```
                        ELSEIF (FInValue(x) <= PWFS(x, y, 2, 0)) THEN
                                MFHIT(x, y) = 255
                        ELSE
                                MFHIT(x, y) = ((PWFS(x, y, 3, 0) - //
FInValue(x)) * 255&) / (PWFS(x, y, 3, 0) - PWFS(x, y, 2, 0))
                        END IF              NEXT y
        END IF
NEXT x
GrndArea = 0
GrndFirstMom = 0
e = 0
DO
    d = 0
    DO
        c = 0
        DO
            b = 0
            DO
                a = 0
                DO
    FuzSum = 255
    IF (MFHIT(4, e) < FuzSum AND NumInputs > 4) THEN
        FuzSum = MFHIT(4, e)
    END IF
    IF (MFHIT(3, d) < FuzSum AND NumInputs > 3) THEN
        FuzSum = MFHIT(3, d)
    END IF
    IF (MFHIT(2, c) < FuzSum AND NumInputs > 2) THEN
        FuzSum = MFHIT(2, c)
    END IF
    IF (MFHIT(1, b) < FuzSum AND NumInputs > 1) THEN
        FuzSum = MFHIT(1, b)
    END IF
    IF (MFHIT(0, a) < FuzSum) THEN
        FuzSum = MFHIT(0, a)
    END IF
    IF (FuzSum <> 0) THEN
        Consequent = Rule(a, b, c, d, e)
        IF (Consequent = -1) THEN
            GOTO Skip
        END IF   'Don't care rule
        FOR I = 0 TO NumOutputs - 1
            IF (SHAPE(I + 5) = 1) THEN   'Output is a singleton
                GrndArea = GrndArea + FuzSum
                GrndFirstMom = GrndFirstMom + FuzSum //
* PWFS(5, Consequent, 0, 0)
            ELSEIF (SHAPE(I + 5) = 3) THEN   'Output is a triangle
                Inter1 = PWFS(5, Consequent, 0, 0) //
+ (FuzSum * (PWFS(5, Consequent, 1, 0) - PWFS(5, Consequent, 0, 0))) / 255&
                Inter2 = PWFS(5, Consequent, 2, 0) //
```

```
 - (FuzSum * (PWFS(5, Consequent, 2, 0) - PWFS(5, Consequent, 1, 0))) / 255&
                    IF (PWFS(5, Consequent, 0, 1) = 255) THEN
                            Area1 = FuzSum * Inter1
                            FirstMoment1 = (Area1 * Inter1) / 2
                            Area2 = FuzSum * (Inter2 - Inter1)
                            FirstMoment2 = Area2 //
 * (Inter2 - (Inter2 - Inter1) / 2)
                    ELSE
                            Area1 = FuzSum //
 * (Inter1 - PWFS(5, Consequent, 0, 0)) / 2
                            FirstMoment1 = Area1 //
 * (Inter1 - ((Inter1 - PWFS(5, Consequent, 0, 0)) / 3))
                            Area2 = FuzSum * (Inter2 - Inter1)
                            FirstMoment2 = Area2 //
 * (Inter1 + (Inter2 - Inter1) / 2)
                    END IF

                    IF (PWFS(5, Consequent, 2, 1) = 255) THEN
                            Area4 = FuzSum //
 * (PWFS(5, Consequent, 2, 0) - Inter2)
                            FirstMoment4 = Area4 //
 * (Inter2 + ((PWFS(5, Consequent, 2, 0) - Inter2) / 2))
                    ELSE
                            Area4 = FuzSum //
 * (PWFS(5, Consequent, 2, 0) - Inter2) / 2
                            FirstMoment4 = Area4 //
 * (Inter2 + ((PWFS(5, Consequent, 2, 0) - Inter2) / 3))
                    END IF
                    Area = Area1 + Area2 + Area4
                    FirstMom = FirstMoment1 + FirstMoment2 + FirstMoment4

                    GrndArea = GrndArea + Area
                    GrndFirstMom = GrndFirstMom + FirstMom
                ELSEIF (SHAPE(I + 5) = 4) THEN   'Is a trapezoid
                    Inter1 = PWFS(5, Consequent, 0, 0) //
 + (FuzSum * (PWFS(5, Consequent, 1, 0) - PWFS(5, Consequent, 0, 0))) / 255
                    Inter2 = PWFS(5, Consequent, 3, 0) //
 - (FuzSum * (PWFS(5, Consequent, 3, 0) - PWFS(5, Consequent, 2, 0))) / 255
                    IF (PWFS(5, Consequent, 0, 1) = 255) THEN
                            Area1 = FuzSum * Inter1
                            FirstMoment1 = (Area1 * Inter1) / 2
                            Area2 = FuzSum * (Inter2 - Inter1)
                            FirstMoment2 = Area2 //
 * (Inter2 - (Inter2 - Inter1) / 2)
                    ELSE
                            Area1 = (FuzSum * (Inter1 //
 - PWFS(5, Consequent, 0, 0))) / 2
                            FirstMoment1 = Area1 //
 * (Inter1 - (Inter1 - PWFS(5, Consequent, 0, 0)) / 3)
                            Area2 = FuzSum * (Inter2 - Inter1)
```

```
                          FirstMoment2 = Area2 //
* (Inter1 + (Inter2 - Inter1) / 2)
                      END IF
                      IF (PWFS(5, Consequent, 3, 1) = 255) THEN
                      Area4 = FuzSum * (PWFS(5, NumFS(5) - 1, 3, 0) - Inter2)
                          FirstMoment4 = Area4 //
* ((PWFS(5, NumFS(5) - 1, 3, 0) - Inter2) / 2)
                      ELSE
                          Area4 = FuzSum //
* (PWFS(5, Consequent, 3, 0) - Inter2) / 2
                          FirstMoment4 = Area4 //
* (Inter2 + (PWFS(5, Consequent, 3, 0) - Inter2) / 3)
                      END IF
                      Area = Area1 + Area2 + Area4
                      FirstMom = FirstMoment1 + FirstMoment2 + FirstMoment4
                      'Summation of pieces of each output shape
                      GrndArea = GrndArea + Area
                      'Summation of different output shapes.
                      GrndFirstMom = GrndFirstMom + FirstMom
                   END IF
             NEXT I
Skip:
      END IF
                          a = a + 1
                      LOOP WHILE (a < NumFS(0))
                      b = b + 1
                  LOOP WHILE (b < NumFS(1) AND NumInputs > 1)
                  c = c + 1
            LOOP WHILE (c < NumFS(2) AND NumInputs > 2)
            d = d + 1
      LOOP WHILE (d < NumFS(3) AND NumInputs > 3)
      e = e + 1
LOOP WHILE (e < NumFS(4) AND NumInputs > 4)
IF (GrndArea <> 0) THEN
      FOutValue(0) = (GrndFirstMom / GrndArea) 'Defuzzify output
ELSE
      FOutValue(0) = 0   'Meaningless output
END IF
END SUB

'Enters data int FInValue(4) as needed by the knowledge base
'NumInputs is input dimension from knowledge base
SUB GetInput
   SCREEN 0, 0      ' Set text screen.
   DO               ' Input titles.
       CLS
       INPUT "Enter Input One: ", FInValue(0)
       IF NumInputs > 1 THEN
          INPUT "Enter Input Two: ", FInValue(1)
       END IF
```

```
        IF NumInputs > 2 THEN
            INPUT "Enter Input Three: ", FInValue(2)
        END IF
        IF NumInputs > 3 THEN
            INPUT "Enter Input Four: ", FInValue(3)
        END IF
        IF NumInputs > 4 THEN
            INPUT "Enter Input Five: ", FInValue(4)
        END IF

        ' Check to see if titles are OK:
        LOCATE 7, 1
        PRINT "OK (Y to continue, N to change)? ";
        LOCATE , , 1
        OK$ = UCASE$(INPUT$(1))
    LOOP UNTIL OK$ = "Y"

END SUB

'Read in knowledge base put out by the Manifold Editor
'Variables are discussed at program start. The variables are global.
SUB ReadKnowledgeBase STATIC

OPEN "TEST.FDT" FOR INPUT AS #1
SEEK #1, 1
DO UNTIL EOF(1)
    INPUT #1, FileString$
    SELECT CASE FileString$
    CASE "NUM_INPUTS"
        INPUT #1, Count$
        NumInputs = VAL(Count$)
    CASE "CREDIBILITY"
        INPUT #1, Count$
        Credibility = VAL(Count$)
    CASE "INPUT1"
        INPUT #1, Count$
        NumFS(0) = VAL(Count$)
    CASE "INPUT2"
        INPUT #1, Count$
        NumFS(1) = VAL(Count$)
    CASE "INPUT3"
        INPUT #1, Count$
        NumFS(2) = VAL(Count$)
    CASE "INPUT4"
        INPUT #1, Count$
        NumFS(3) = VAL(Count$)
    CASE "INPUT5"
        INPUT #1, Count$
        NumFS(4) = VAL(Count$)
    CASE "NUM_OUTPUTS"
```

```
      INPUT #1, Count$
      NumOutputs = VAL(Count$)
CASE "OUTPUT1"
      INPUT #1, Count$
      NumFS(5) = VAL(Count$)
CASE "OUTPUT2"
      INPUT #1, Count$
      NumFS(6) = VAL(Count$)
CASE "NUM_RULES"
      INPUT #1, Count$
      NumRules = VAL(Count$)
      PRINT NumRules
CASE "INPUTS FUZZY SETS"
      FOR I = 0 TO NumInputs - 1
           DO
           INPUT #1, FileString$
           SELECT CASE FileString$
           CASE "INPUT"
                INPUT #1, Count$
                M = 0
           CASE "COUNT"
                INPUT #1, Count$
                IF NumFS(I) <> VAL(Count$) THEN
                     END
                END IF
           CASE "SHAPE"
                INPUT #1, Count$
                SHAPE(I) = VAL(Count$)
           CASE "START"
                FOR J = 0 TO NumFS(I) - 1
                     FOR K = 0 TO SHAPE(I) - 1
                          INPUT #1, Count$
                          PRINT Count$
                          PWFS(I, J, K, 0) = VAL(Count$)
                          PRINT PWFS(I, J, K, 0)
                          INPUT #1, Count$
                          PRINT Count$
                          PWFS(I, J, K, 1) = VAL(Count$)
                          PRINT PWFS(I, J, K, 1)
                     NEXT K
                NEXT J
           END SELECT
           LOOP WHILE FileString$ <> "END$"
      NEXT I
CASE "OUTPUTS FUZZY SETS"
      FOR I = 0 TO NumOutputs - 1
           DO
           INPUT #1, FileString$
           SELECT CASE FileString$
           CASE "OUTPUT"
```

```
                    INPUT #1, Count$
                    M = 5
              CASE "COUNT"
                    INPUT #1, Count$
                    IF NumFS(I + M) <> VAL(Count$) THEN
                          END
                    END IF
              CASE "SHAPE"
                    INPUT #1, Count$
                    SHAPE(I + M) = VAL(Count$)
              CASE "START"
                    FOR J = 0 TO NumFS(I + M) - 1
                          FOR K = 0 TO SHAPE(I + M) - 1
                                INPUT #1, Count$
                                PRINT Count$
                                PWFS(I + M, J, K, 0) = VAL(Count$)
                                IF (SHAPE(I + M) > 1) THEN
                                      INPUT #1, Count$
                                      PRINT Count$
                                      PWFS(I + M, J, K, 1) = VAL(Count$)
                                END IF
                          NEXT K
                    NEXT J
              'CASE "END$"
              'CASE ELSE

              END SELECT
              LOOP WHILE FileString$ <> "END$"
        NEXT I
    CASE "RULES"
        FOR H = 0 TO 4
            IF NumFS(H) = 0 THEN Temp(H) = 0 ELSE Temp(H) = NumFS(H) - 1
        NEXT H
        FOR I = 0 TO Temp(0)
            FOR J = 0 TO Temp(1)
                FOR K = 0 TO Temp(2)
                    FOR L = 0 TO Temp(3)
                        FOR M = 0 TO Temp(4)
                            INPUT #1, Rule(I, J, K, L, M)
                            PRINT Rule(I, J, K, L, M)
                        NEXT M
                    NEXT L
                NEXT K
            NEXT J
        NEXT I
    CASE ELSE

    END SELECT
LOOP CLOSE #1
END SUB
```

```
SUB ShowOutput
    SCREEN 0, 0        ' Set text screen.
    DO               ' Input titles.
        CLS
        PRINT "Output One: "; FOutValue(0)
        IF NumOutputs > 1 THEN PRINT "Output Two: "; FOutValue(1)
        LOCATE 7, 1
        PRINT "OK (Y to continue)? ";
        LOCATE , , 1
        OK$ = UCASE$(INPUT$(1))
    LOOP UNTIL OK$ = "Y"
END SUB
```

QUICKBASIC FAST INFERENCE ENGINE

This inference engine is *fuzzy2.bas,* and can also be be tested with the file *test.fdt* that's on the disk or with any file you name *test.fdt.* You can also substitute any other file name with the .fdt extension. In that case, you should change the references in fuzzy2.bas. These are in bold type in the program listing.

Like fuzzy1.fdt, fuzzy2.bas can handle triangular and trapezoidal membership functions. It's faster because tests only the active rules.

The code for fuzzy2.bas follows. (To accommodate the printed page, long lines that must carry over are broken with a double slash, / /.)

```
'Fuzzy Exersizer
'Reads TEST.FDT
'Allows
DECLARE SUB ShowOutput ()
DECLARE SUB ReadKnowledgeBase ()
DECLARE SUB FuzzyMap ()
DECLARE SUB GetInput ()
DECLARE SUB DisplayOutput ()
COMMON Rule() AS INTEGER
'Global. From XXXX.FDT.
'The rule array size is defined at knowledge base read time.
DIM SHARED NumInputs AS INTEGER
'Global. From XXXX.FDT.
'The Number of Inputs or input dimensions(1 to 5)
DIM SHARED NumOutputs AS INTEGER
'The Number of Outputs (1 to 2) Each output implies a separate estimation
'surface. The input permutations are the same.
```

```
DIM SHARED Credibility AS INTEGER
'Ignore Credibility now. The inference engine does not use it.
DIM SHARED NumRules AS INTEGER
'Global. From XXXX.FDT.
'The number of rules to be read in.
DIM SHARED NumFS(6) AS INTEGER
'Global. From XXXX.FDT.
'The number of fuzzy sets on each input (0 to 4) and each output (5 to 6)
'This may vary from 1 to 11 on each input or output.
'Used to control readin and inference engine.
DIM SHARED PWFS(6, 10, 3, 1) AS INTEGER
'Global. From XXXX.FDT.
'The piecewise fuzzy sets are stored here:
'0 to 6: The input and output indicies, same as NumFS(6)
'0 to 10: The fuzzy set index on each input or output dimension.
'    The maximum value is the same as stored in NumFS(6). '0 to 3: The num-
ber of points on each fuzzy set in each dimension.
'    The values may be: 1 = Singleton, 3 = Triangle, 4 = trapezoid.
'    Same as value stored in SHAPE(6).
'0 to 1: The number of values needed to define each vertex point of the
'    fuzzy sets. 1 = Singleton, 2 = Other.
DIM SHARED FInValue(4) AS INTEGER
'Global.
'Values into the inference engine. From GetInput sub or from
'    your application.
DIM SHARED FOutValue(1) AS INTEGER
'Global.
'Values from the inference engine. To ShowOutput sub or to
'    your application.
DIM SHARED SHAPE(6) AS INTEGER
'Global. From XXXX.FDT.
'    The values may be: 1 = Singleton, 3 = Triangle, 4 = trapezoid.
DEFINT A-Z

DO
    CLS
    PRINT "Fuzzy Work": PRINT
    COLOR 15, 0: PRINT "  R"; :   COLOR 7, 0: PRINT "ead Knowledge Base"
    COLOR 15, 0: PRINT "  F"; :   COLOR 7, 0: PRINT "uzzy Map"
    COLOR 15, 0: PRINT "  I"; :   COLOR 7, 0: PRINT "nput Values"
    COLOR 15, 0: PRINT "  S"; :   COLOR 7, 0: PRINT "how Output Values"
    COLOR 15, 0: PRINT "  Q"; :   COLOR 7, 0: PRINT "uit"
    PRINT : PRINT "Select: "

    ' Get valid key
    DO
        Q$ = UCASE$(INPUT$(1))
    LOOP WHILE INSTR("BRFISQ", Q$) = 0

    ' Take action based on key
```

```
CLS
SELECT CASE Q$
    CASE IS = "R"
        PRINT "Reading . . ."
        OPEN "TEST.FDT" FOR INPUT AS #1
        DO UNTIL EOF(1)
            INPUT #1, FileString$
            SELECT CASE FileString$
            CASE "NUM_INPUTS"
                INPUT #1, Count$
                NumInputs = VAL(Count$)
            CASE "CREDIBILITY"
                INPUT #1, Count$
                Credibility = VAL(Count$)
            CASE "INPUT1"
                INPUT #1, Count$
                NumFS(0) = VAL(Count$)
            CASE "INPUT2"
                INPUT #1, Count$
                NumFS(1) = VAL(Count$)
            CASE "INPUT3"
                INPUT #1, Count$
                NumFS(2) = VAL(Count$)
            CASE "INPUT4"
                INPUT #1, Count$
                NumFS(3) = VAL(Count$)
            CASE "INPUT5"
                INPUT #1, Count$
                NumFS(4) = VAL(Count$)
            CASE "NUM_OUTPUTS"
                INPUT #1, Count$
                NumOutputs = VAL(Count$)
            CASE "OUTPUT1"
                INPUT #1, Count$
                NumFS(5) = VAL(Count$)
            CASE "OUTPUT2"
                INPUT #1, Count$
                NumFS(6) = VAL(Count$)
            CASE "NUM_RULES"
                INPUT #1, Count$
                NumRules = VAL(Count$)
            CASE ELSE

            END SELECT
        LOOP
        CLOSE #1
        REDIM SHARED Rule(NumFS(0), NumFS(1), NumFS(2), NumFS(3), //
NumFS(4)) AS INTEGER
        ReadKnowledgeBase
        PRINT "Fuzzy Mapping . . ."
```

```
        CASE IS = "F"
            PRINT "Fuzzy Mapping . . ."
          FuzzyMap
          ShowOutput
        CASE IS = "I"
            PRINT "Inputing . . ."
            GetInput
        CASE IS = "S"
            PRINT "Outputing . . ."
            ShowOutput          CASE ELSE
    END SELECT
LOOP UNTIL Q$ = "Q"
END


'Display the inferred outputs previously calculated.
SUB DisplayOutput
    SCREEN 0, 0       ' Set text screen.
    DO              ' Input titles.
        CLS
        LOCATE 7, 1
        PRINT "Output 1: "; FOutValue(0)
        IF NumInputs > 1 THEN
            PRINT "Output 2: "; FOutValue(1)
        END IF
        PRINT "OK (Y to continue, N to change)? ";
        LOCATE , , 1
        OK$ = UCASE$(INPUT$(1))
    LOOP UNTIL OK$ = "Y"

END SUB

'Inference Engine.
'Up to 5 inputs and 1 output supported
'The crisp input value on each input dimension is tested for nonzero value.
'If a nonzero value occurs, it's value and which fuzzy set it is in is
'registered.
'Then all the rules with all nonzero fuzzy sets are "fired" to provide
'the associated output fuzzy set and the minimum input hit value is used to
'truncate the output fuzzy set. The truncated output fuzzy sets areas are
'summed and the center of gravity of each of the primative pieces times
'the primative area(first moment) is summed.
'Primative areas are triangles and rectangles.
'Then the area sumation results is divided into the first moment summation
'to give the overall center of gravity. This is the crisp output of
'inference process.
SUB FuzzyMap
DIM GrndArea AS LONG, GrndFirstMom AS LONG
'Local.
'Accumulation variables for output pieces.
DIM FuzSum AS LONG
```

```
'Local. 'The minimum nonzero membership value of a rule that has fired.
DIM MFHIT(0 TO 4, 0 TO 10) AS LONG
'Local.
'The membership values of the crisp inputs in each fuzzy set in each input
'dimension. Most are zero.
DIM Consequent AS LONG
'Local.
'The rule conseguent for the particular rule that is firing.
'Firing means all fuzzified inputs are nonzero on this rule.
DIM Area1 AS LONG, FirstMoment1 AS LONG
'Local.
'Left hand triangle shape area of output truncated trapezoid.
'This shape area times the Center of Gravity (First Moment) of this shape.
DIM Area2 AS LONG, FirstMoment2 AS LONG
'Local.
'Center rectangle shape area of output truncated trapezoid.
'This shape area times the Center of Gravity (First Moment) of this shape.
DIM Area4 AS LONG, FirstMoment4 AS LONG
'Local.
'Right hand triangle shape area of output truncated trapezoid.
'This shape times the Center of Gravity (First Moment) of this shape.
DIM Area AS LONG, FirstMom AS LONG
'Local.
'Intermediate sumation variables. May not be needed.
DIM Inter1 AS LONG, Inter2 AS LONG
'Local.
'Used in the defuzzification process. Stores the two output dimension values
'  of the trapezoidal peak after truncation. (Triangles turn into trapezoids
'  after truncation.
DIM HitsonDim(4) AS LONG
'Local.
'Used to register the number of nonzero input hits in each input dimension.
DIM Tmp(4) AS LONG
'Local.
'Used to setup the FOR loop indicies on the number of nonzero
'input hits in each input dimension.
DIM FSHit(4, 10) AS LONG
'Local.
'Used to register the input dimension and fuzzy set in which the fuzzified
'input is nonzero.

' COMPUTE Hit Values on memberships */ FOR H = 0 TO 4
    HitsonDim(H) = 0    'Start out with all hit count = 0
NEXT H

FOR x = 0 TO NumInputs - 1
    IF (SHAPE(x) = 3) THEN    'Trangles
        z = 0
        FOR y = 0 TO NumFS(x) - 1
            IF (FInValue(x) <= PWFS(x, y, 0, 0) OR FInValue(x) //
```

```
                    >= PWFS(x, y, 2, 0)) THEN
                                    'MFHIT(x, y) = 0  'Missed fuzzy set entirely
                            ELSEIF (FInValue(x) <= PWFS(x, 0, 1, 0) AND //
PWFS(x, 0, 0, 1) = 255) THEN
                                    HitsonDim(x) = HitsonDim(x) + 1
                                    FSHit(x, z) = y
                                    MFHIT(x, z) = 255  'Hit high left end
                                    z = z + 1
                            ELSEIF (FInValue(x) >= PWFS(x, NumFS(x) - 1, 1, 0) //
AND PWFS(x, NumFS(x) - 1, 2, 1) = 255) THEN
                                    HitsonDim(x) = HitsonDim(x) + 1
                                    FSHit(x, z) = y
                                    MFHIT(x, z) = 255  'Hit high right end
                                    z = z + 1
                            ELSEIF (FInValue(x) <= PWFS(x, y, 1, 0)) THEN
                                    HitsonDim(x) = HitsonDim(x) + 1
                                    FSHit(x, z) = y
                                    MFHIT(x, z) = ((FInValue(x) - PWFS(x, y, 0, 0)) * 255&)
/ (PWFS(x, y, 1, 0) - PWFS(x, y, 0, 0))
                                    'Hit is on left slope and between 0 and 255
                                    z = z + 1
                            ELSEIF (FInValue(x) > PWFS(x, y, 1, 0)) THEN
                                    HitsonDim(x) = HitsonDim(x) + 1
                                    FSHit(x, z) = y
                                    MFHIT(x, z) = ((PWFS(x, y, 2, 0) //
- FInValue(x)) * 255&) / (PWFS(x, y, 2, 0) - PWFS(x, y, 1, 0))
                                    'Hit is on right slope and between 0 and 255
                                    z = z + 1
                            END IF
                    NEXT y
            ELSEIF (SHAPE(x) = 4) THEN 'Trapezoid
                    z = 0
                    FOR y = 0 TO NumFS(x) - 1
                            IF (FInValue(x) <= PWFS(x, y, 0, 0) OR FInValue(x) //
>= PWFS(x, y, 3, 0)) THEN
                                    MFHIT(x, y) = 0
                            ELSEIF (FInValue(x) <= PWFS(x, 0, 1, 0) //
AND PWFS(x, 0, 0, 1) = 255) THEN
                                    HitsonDim(x) = HitsonDim(x) + 1
                                    FSHit(x, z) = y
                                    MFHIT(x, z) = 255                    z = z + 1
                            ELSEIF (FInValue(x) >= PWFS(x, NumFS(x) - 1, 1, 0) //
AND PWFS(x, NumFS(x) - 1, 3, 1) = 255) THEN
                                    HitsonDim(x) = HitsonDim(x) + 1
                                    FSHit(x, z) = y
                                    MFHIT(x, z) = 255
                                    z = z + 1
                            ELSEIF (FInValue(x) <= PWFS(x, y, 1, 0)) THEN
                                    HitsonDim(x) = HitsonDim(x) + 1
                                    FSHit(x, z) = y
```

```
                        MFHIT(x, z) = ((FInValue(x) - PWFS(x, y, 0, 0)) //
* 255&) / (PWFS(x, y, 1, 0) - PWFS(x, y, 0, 0))
                        z = z + 1
                ELSEIF (FInValue(x) <= PWFS(x, y, 2, 0)) THEN
                        HitsonDim(x) = HitsonDim(x) + 1
                        FSHit(x, z) = y
                        MFHIT(x, z) = 255
                        z = z + 1
                ELSE
                        HitsonDim(x) = HitsonDim(x) + 1
                        FSHit(x, z) = y
                        MFHIT(x, z) = ((PWFS(x, y, 3, 0) //
- FInValue(x)) * 255&) / (PWFS(x, y, 3, 0) - PWFS(x, y, 2, 0))
                        z = z + 1
                END IF
            NEXT y
        END IF
NEXT x
GrndArea = 0
GrndFirstMom = 0
FOR H = 0 TO 4
    IF HitsonDim(H) = 0 THEN Tmp(H) = 0 ELSE Tmp(H) = HitsonDim(H) - 1
NEXT H
FOR e = 0 TO Tmp(4)
    FOR d = 0 TO Tmp(3)
        FOR c = 0 TO Tmp(2)
            FOR b = 0 TO Tmp(1)
                FOR a = 0 TO Tmp(0)
    FuzSum = 255
    IF (MFHIT(4, e) < FuzSum AND HitsonDim(4) > 0) THEN
        FuzSum = MFHIT(4, e)
    END IF
    IF (MFHIT(3, d) < FuzSum AND HitsonDim(3) > 0) THEN
        FuzSum = MFHIT(3, d)
    END IF
    IF (MFHIT(2, c) < FuzSum AND HitsonDim(2) > 0) THEN
        FuzSum = MFHIT(2, c)
    END IF
    IF (MFHIT(1, b) < FuzSum AND HitsonDim(1) > 0) THEN
        FuzSum = MFHIT(1, b)
    END IF
    IF (MFHIT(0, a) < FuzSum AND HitsonDim(0) > 0) THEN
        FuzSum = MFHIT(0, a)
    END IF
    Consequent = Rule(FSHit(0, a), FSHit(1, b), FSHit(2, c), //
FSHit(3, d), FSHit(4, e))
    IF (Consequent = -1) THEN
        GOTO Skip   'Don't care rule
    END IF
        IF (SHAPE(I + 5) = 1) THEN   'Output is a singleton
```

```
                        GrndArea = GrndArea + FuzSum
                        GrndFirstMom = GrndFirstMom + FuzSum //
* PWFS(5, Consequent, 0, 0)
              ELSEIF (SHAPE(I + 5) = 3) THEN   'Output is a triangle
                        Inter1 = PWFS(5, Consequent, 0, 0) + (FuzSum //
* (PWFS(5, Consequent, 1, 0) - PWFS(5, Consequent, 0, 0))) / 255&
                        Inter2 = PWFS(5, Consequent, 2, 0) - (FuzSum //
* (PWFS(5, Consequent, 2, 0) - PWFS(5, Consequent, 1, 0))) / 255&
                  IF (PWFS(5, Consequent, 0, 1) = 255) THEN
                        Area1 = FuzSum * Inter1
                        FirstMoment1 = (Area1 * Inter1) / 2
                        Area2 = FuzSum * (Inter2 - Inter1)
                        FirstMoment2 = Area2 * (Inter2 - (Inter2 - Inter1) / 2)
                  ELSE
                        Area1 = FuzSum * (Inter1 - PWFS(5, Consequent, 0, 0)) //
/ 2
                        FirstMoment1 = Area1 * (Inter1 - (Inter1 //
- PWFS(5, Consequent, 0, 0)) / 3)
                        Area2 = FuzSum * (Inter2 - Inter1)
                        FirstMoment2 = Area2 * (Inter1 + (Inter2 - Inter1) / 2)
                  END IF
                  IF (PWFS(5, Consequent, 2, 1) = 255) THEN
                        Area4 = FuzSum * (PWFS(5, Consequent, 2, 0) - Inter2)
                        FirstMoment4 = Area4 * (Inter2 //
+ (PWFS(5, Consequent, 2, 0) - Inter2) / 2)
                  ELSE
                        Area4 = FuzSum * (PWFS(5, Consequent, 2, 0) //
- Inter2) / 2
                        FirstMoment4 = Area4 * (Inter2 //
+ (PWFS(5, Consequent, 2, 0) - Inter2) / 3)
                  END IF
                  Area = Area1 + Area2 + Area4
                  FirstMom = FirstMoment1 + FirstMoment2 + FirstMoment4

                  GrndArea = GrndArea + Area
                  GrndFirstMom = GrndFirstMom + FirstMom
              ELSEIF (SHAPE(I + 5) = 4) THEN   'Is a trapezoid
                        Inter1 = PWFS(5, Consequent, 0, 0) + (FuzSum //
* (PWFS(5, Consequent, 1, 0) - PWFS(5, Consequent, 0, 0))) / 255
                        Inter2 = PWFS(5, Consequent, 3, 0) - (FuzSum //
* (PWFS(5, Consequent, 3, 0) - PWFS(5, Consequent, 2, 0))) / 255
                  IF (PWFS(5, Consequent, 0, 1) = 255) THEN
                        Area1 = FuzSum * Inter1
                        FirstMoment1 = (Area1 * Inter1) / 2
                        Area2 = FuzSum * (Inter2 - Inter1)
                        FirstMoment2 = Area2 * (Inter2 - (Inter2 - Inter1) / 2)
                  ELSE
                        Area1 = (FuzSum * (Inter1 //
- PWFS(5, Consequent, 0, 0))) / 2
                        FirstMoment1 = Area1 * (Inter1 //
```

```
- (Inter1 - PWFS(5, Consequent, 0, 0)) / 3)
                    Area2 = FuzSum * (Inter2 - Inter1)
                    FirstMoment2 = Area2 * (Inter1 + (Inter2 - Inter1) / 2)
                END IF
                IF (PWFS(5, Consequent, 3, 1) = 255) THEN
                    Area4 = FuzSum * (PWFS(5, NumFS(5) - 1, 3, 0) - Inter2)
                    FirstMoment4 = Area4 //
* ((PWFS(5, Consequent, 3, 0) - Inter2) / 2)
                ELSE
                    Area4 = FuzSum //
* (PWFS(5, Consequent, 3, 0) - Inter2) / 2
                    FirstMoment4 = Area4 //
* (Inter2 + (PWFS(5, Consequent, 3, 0) - Inter2) / 3)
                END IF

                Area = Area1 + Area2 + Area4
                FirstMom = FirstMoment1 + FirstMoment2 + FirstMoment4
                'Summation of pieces of each output shape
                GrndArea = GrndArea + Area
                'Summation of different output shapes.
                GrndFirstMom = GrndFirstMom + FirstMom
            END IF
Skip:
                    NEXT a
                NEXT b
            NEXT c
        NEXT d
NEXT e
IF (GrndArea <> 0) THEN
    FOutValue(0) = (GrndFirstMom / GrndArea) 'Defuzzify output
ELSE      FOutValue(0) = 0   'Meaningless output
END IF
END SUB

'Enters data int FInValue(4) as needed by the knowledge base
'NumInputs is input dimension from knowledge base
SUB GetInput
   SCREEN 0, 0       ' Set text screen.
   DO               ' Input titles.
      CLS
      INPUT "Enter Input One: ", FInValue(0)
      IF NumInputs > 1 THEN
         INPUT "Enter Input Two: ", FInValue(1)
      END IF
      IF NumInputs > 2 THEN
         INPUT "Enter Input Three: ", FInValue(2)
      END IF
      IF NumInputs > 3 THEN
         INPUT "Enter Input Four: ", FInValue(3)
      END IF
```

```
        IF NumInputs > 4 THEN
            INPUT "Enter Input Five: ", FInValue(4)
        END IF

        ' Check to see if titles are OK:
        LOCATE 7, 1
        PRINT "OK (Y to continue, N to change)? ";
        LOCATE , , 1
        OK$ = UCASE$(INPUT$(1))
    LOOP UNTIL OK$ = "Y"

END SUB

'Read in knowledge base put out by the Manifold Editor
'Variables are discussed at program start. The variables are global.
SUB ReadKnowledgeBase STATIC

OPEN "TEST.FDT" FOR INPUT AS #1
SEEK #1, 1
DO UNTIL EOF(1)
    INPUT #1, FileString$
    SELECT CASE FileString$
    CASE "NUM_INPUTS"
        INPUT #1, Count$
        NumInputs = VAL(Count$)
    CASE "CREDIBILITY"
        INPUT #1, Count$
        Credibility = VAL(Count$)
    CASE "INPUT1"            INPUT #1, Count$
        NumFS(0) = VAL(Count$)
    CASE "INPUT2"
        INPUT #1, Count$
        NumFS(1) = VAL(Count$)
    CASE "INPUT3"
        INPUT #1, Count$
        NumFS(2) = VAL(Count$)
    CASE "INPUT4"
        INPUT #1, Count$
        NumFS(3) = VAL(Count$)
    CASE "INPUT5"
        INPUT #1, Count$
        NumFS(4) = VAL(Count$)
    CASE "NUM_OUTPUTS"
        INPUT #1, Count$
        NumOutputs = VAL(Count$)
    CASE "OUTPUT1"
        INPUT #1, Count$
        NumFS(5) = VAL(Count$)
    CASE "OUTPUT2"
        INPUT #1, Count$
```

```
            NumFS(6) = VAL(Count$)
      CASE "NUM_RULES"
            INPUT #1, Count$
            NumRules = VAL(Count$)
            PRINT NumRules
      CASE "INPUTS FUZZY SETS"
            FOR I = 0 TO NumInputs - 1
                  DO
                  INPUT #1, FileString$
                  SELECT CASE FileString$
                  CASE "INPUT"
                        INPUT #1, Count$
                        M = 0
                  CASE "COUNT"
                        INPUT #1, Count$
                        IF NumFS(I) <> VAL(Count$) THEN
                              END
                        END IF
                  CASE "SHAPE"
                        INPUT #1, Count$
                        SHAPE(I) = VAL(Count$)
                  CASE "START"
                        FOR J = 0 TO NumFS(I) - 1
                              FOR K = 0 TO SHAPE(I) - 1
                                    INPUT #1, Count$
                                    PRINT Count$
                                    PWFS(I, J, K, 0) = VAL(Count$)
                                    INPUT #1, Count$
                                    PRINT Count$
                                    PWFS(I, J, K, 1) = VAL(Count$)
                              NEXT K
                        NEXT J
                  END SELECT
                  LOOP WHILE FileString$ <> "END$"
            NEXT I
      CASE "OUTPUTS FUZZY SETS"
            FOR I = 0 TO NumOutputs - 1
                  DO
                  INPUT #1, FileString$
                  SELECT CASE FileString$
                  CASE "OUTPUT"
                        INPUT #1, Count$
                        M = 5
                  CASE "COUNT"
                        INPUT #1, Count$
                        IF NumFS(I + M) <> VAL(Count$) THEN
                              END
                        END IF
                  CASE "SHAPE"
                        INPUT #1, Count$
```

```
                              SHAPE(I + M) = VAL(Count$)
               CASE "START"
                    FOR J = 0 TO NumFS(I + M) - 1
                         FOR K = 0 TO SHAPE(I + M) - 1
                              INPUT #1, Count$
                              PRINT Count$
                              PWFS(I + M, J, K, 0) = VAL(Count$)
                              IF (SHAPE(I + M) > 1) THEN
                                   INPUT #1, Count$
                                   PRINT Count$
                                   PWFS(I + M, J, K, 1) = VAL(Count$)
                              END IF
                         NEXT K
                    NEXT J
               'CASE "END$"
               'CASE ELSE

               END SELECT
               LOOP WHILE FileString$ <> "END$"
          NEXT I
     CASE "RULES"
          FOR H = 0 TO 4
               IF NumFS(H) = 0 THEN Temp(H) = 0 ELSE Temp(H) = NumFS(H) - 1
          NEXT H
          FOR I = 0 TO Temp(0)
               FOR J = 0 TO Temp(1)                         FOR K = 0 TO Temp(2)
                    FOR L = 0 TO Temp(3)
                         FOR M = 0 TO Temp(4)
                              INPUT #1, Rule(I, J, K, L, M)
                              PRINT Rule(I, J, K, L, M)
                         NEXT M
                    NEXT L
               NEXT K
          NEXT J
     NEXT I
     CASE ELSE

     END SELECT
LOOP
CLOSE #1
END SUB


SUB ShowOutput
   SCREEN 0, 0      ' Set text screen.
   DO               ' Input titles.
      CLS
      PRINT "Output One: "; FOutValue(0)
      IF NumOutputs > 1 THEN PRINT "Output Two: "; FOutValue(1)
      LOCATE 7, 1
      PRINT "OK (Y to continue)? ";
```

```
        LOCATE , , 1
        OK$ = UCASE$(INPUT$(1))
    LOOP UNTIL OK$ = "Y"
END SUB
```

C LANGUAGE INFERENCE ENGINE

The disk file **cie.c** contains code fragments of a general purpose inference engine written in C language. The knowledge base file produced for it by the Fuzzy Knowledge Builder™, also on the disk, is named *testie.fic*. The files are included as clues for those interested in writing a C inference engine.

This inference engine is identical to Fuzzy1, except that it won't handle trapezoidal membership functions. It's a general purpose engine because the switching code included in the fragments quickly switches the specific knowledge base into the general variables used in the inference engine. The inference engine can quickly be switched among knowledge bases.

In the case of testie.fic, the general structure is named IETag and the specific structure is TESTIEInfo.

The most important part of the code follows. (Line irregularities are due to printed page requirements.)

```
// This is a good general purpose inference engine. It is simple and very
slow.
// It will only handle triangular shaped fuzzy sets.
BOOL BLD_FuzzyMapUDCFunc(HWND hWnd,UINT message,WPARAM wParam,LPARAM lParam)
    {
    int a, b, c, d, e, f, g, h, i, j, k, x, y;
    double  dbGrndArea;
    double  dbGrndFirstMom;
    double  dbFuzSum;
    double  dbMFHit[11][11];
    DWORD dwConsequent;
    double dbArea, dbFirstMom;
    double dbArea_1,  dbArea_2, dbArea_3, dbArea_4;
    double dbFirstMoment_1, dbFirstMoment_2, dbFirstMoment_3,
dbFirstMoment_4;
    double dbInter_1, dbInter_2;
    DWORD      dwA[12];
    DWORD  iOffset;
          /* COMPUTE input membership */
    for (x = 0; x <= fie.iFamInDimMax; x++){
```

```
        for (y = 0 ; y <= fie.iFamSecMax[x];y++){
            if (dbIEIn[x] <= dbPWMF[x][y][0][0]
                || dbIEIn[x] >= dbPWMF[x][y][2][0]){
                dbMFHit[x][y] = 0.;
                }
            else {
                if (dbIEIn[x] <= dbPWMF[x][y][1][0]) {
                    dbMFHit[x][y] = ((dbIEIn[x] - dbPWMF[x][y][0][0]))
                                            /(dbPWMF[x][y][1][0] -
dbPWMF[x][y][0][0]);
                    }
                else {
                    dbMFHit[x][y] = ((dbPWMF[x][y][2][0] - dbIEIn[x]))
                                            /(dbPWMF[x][y][2][0] -
dbPWMF[x][y][1][0]);
                    }
                }
            }
        if (dbIEIn[x] <= dbPWMF[x][0][1][0]
                    && dbPWMF[x][0][0][1] == 1.){
            dbMFHit[x][0] = 1.;
            }
        if (dbIEIn[x] >= dbPWMF[x][fie.iFamSecMax[x]][1][0]
                    && dbPWMF[x][fie.iFamSecMax[x]][2][1] == 1.){
            dbMFHit[x][fie.iFamSecMax[x]] = 1.;
            }
        }

        /* Rules */
    for (x = 0; x < MAXINDIM; x++)
        dwA[x] = 0;
    dwA[fie.iFamInDimMax] = 1;
    for (x = fie.iFamInDimMax - 1; x >= 0; x--)
        {
        dwA[x] = dwA[x + 1] * (DWORD)(fie.iFamSecMax[x + 1] + 1);
        }
    dbGrndArea = 0.;
    dbGrndFirstMom = 0.;
        k = 0;
        do
            {
            j = 0;
            do
                {
    i = 0;
    do
        {
        h = 0;
        do
            {
```

```
                       g = 0;
                       do
                               {
          f = 0;
          do
               {                e = 0;
               do
                       {
                       d = 0;
                       do
                               {
          c = 0;
          do
               {
               b = 0;
               do
                       {
                       a = 0;
                       do
                               {
                               dbFuzSum = min(dbMFHit[0][a],GRIDYMAX);
                               if (fie.iFamInDimMax > 0) dbFuzSum =
min(dbMFHit[1][b],dbFuzSum);
                               if (fie.iFamInDimMax > 1)    dbFuzSum =
min(dbMFHit[2][c],dbFuzSum);
                               if (fie.iFamInDimMax > 2)    dbFuzSum =
min(dbMFHit[3][d],dbFuzSum);
                               if (fie.iFamInDimMax > 3)    dbFuzSum =
min(dbMFHit[4][e],dbFuzSum);
                               if (fie.iFamInDimMax > 4)    dbFuzSum =
min(dbMFHit[5][f],dbFuzSum);
                               if (fie.iFamInDimMax > 5)    dbFuzSum =
min(dbMFHit[6][g],dbFuzSum);
                               if (fie.iFamInDimMax > 6)    dbFuzSum =
min(dbMFHit[7][h],dbFuzSum);
                               if (fie.iFamInDimMax > 7)    dbFuzSum =
min(dbMFHit[8][i],dbFuzSum);
                               if (fie.iFamInDimMax > 8)    dbFuzSum =
min(dbMFHit[9][j],dbFuzSum);
                               if (fie.iFamInDimMax > 9)    dbFuzSum =
min(dbMFHit[10][k],dbFuzSum);
                               if (dbFuzSum != 0){
                                   iOffset = (((DWORD)k * dwA[10]) +
                                            ((DWORD)j * dwA[9]) +
                                            ((DWORD)i * dwA[8]) +
                                            ((DWORD)h * dwA[7]) +
                                            ((DWORD)g * dwA[6]) +
                                            ((DWORD)f * dwA[5]) +
                                            ((DWORD)e * dwA[4]) +
                                            ((DWORD)d * dwA[3]) +
```

```
                                        ((DWORD)c * dwA[2]) +
                                        ((DWORD)b * dwA[1]) +
                                        ((DWORD)a * dwA[0])));
                dwConsequent = (DWORD)(int)*(gaiRules + iOffset);
                dbInter_1 = dbPWMF[11][dwConsequent][0][0]
                                        + (dbFuzSum *
(dbPWMF[11][dwConsequent][1][0]
- dbPWMF[11][dwConsequent][0][0]));
                dbInter_2 = dbPWMF[11][dwConsequent][2][0]
                                        - (dbFuzSum *
(dbPWMF[11][dwConsequent][2][0]
- dbPWMF[11][dwConsequent][1][0]));
                if (dbPWMF[11][dwConsequent][0][1] == 1.){
dbArea_1 = dbFuzSum * (dbInter_1 - fie.dbFamDimUnitMin[11]);
                        dbFirstMoment_1 = dbArea_1 * (dbInter_1 -
fie.dbFamDimUnitMin[11])/2;
                        dbArea_2 = dbFuzSum * (dbInter_2 - dbInter_1 -
fie.dbFamDimUnitMin[11]);
                        dbFirstMoment_2 = dbArea_2
* (dbInter_2 - fie.dbFamDimUnitMin[11]
- (dbInter_2 - dbInter_1)/2);
                        }
                else {
                        dbArea_1 = (dbFuzSum * (dbInter_1 -
dbPWMF[11][dwConsequent][0][0]))/2;
                        dbFirstMoment_1 = dbArea_1
* (dbInter_1 - fie.dbFamDimUnitMin[11]
                                            - (dbInter_1 -
dbPWMF[11][dwConsequent][0][0])/3);
                        dbArea_2 = dbFuzSum * (dbInter_2 - dbInter_1);
                        dbFirstMoment_2 = dbArea_2 * (dbInter_1 -
fie.dbFamDimUnitMin[11] + (dbInter_2 - dbInter_1)/2);
                        }
                if (dbPWMF[11][dwConsequent][2][1] == 1.){
dbArea_4 = dbFuzSum * (fie.dbFamDimUnitMax[11] - dbInter_2);
                        dbFirstMoment_4 = dbArea_4 * (fie.dbFamDimU-
nitMax[11] - ((fie.dbFamDimUnitMax[11] - dbInter_2)/2));
                        }
                else {                                        dbArea_4 =
dbFuzSum * (dbPWMF[11][dwConsequent][2][0] - dbInter_2)/2;
                        dbFirstMoment_4 = dbArea_4 * (dbInter_2 -
fie.dbFamDimUnitMin[11] + (dbPWMF[11][dwConsequent][2][0] - dbInter_2)/3);
                        }
                dbArea = dbArea_1
                                + dbArea_2
                                + dbArea_4;
                dbFirstMom = dbFirstMoment_1
                                + dbFirstMoment_2
                                + dbFirstMoment_4;
```

```
                            dbGrndArea += dbArea;
                            dbGrndFirstMom += dbFirstMom;
                            }
                       }
               while (++a <= fie.iFamSecMax[0]) ;
               }
          while (++b <= fie.iFamSecMax[1]) ;
          }
   while (++c <= fie.iFamSecMax[2]) ;
                       }
               while (++d <= fie.iFamSecMax[3]) ;
               }
          while (++e <= fie.iFamSecMax[4]) ;
          }
   while (++f <= fie.iFamSecMax[5]) ;
                       }
               while (++g <= fie.iFamSecMax[6]) ;
               }
          while (++h <= fie.iFamSecMax[7]) ;
          }
   while (++i <= fie.iFamSecMax[8]) ;
                       }
               while (++j <= fie.iFamSecMax[9]) ;
               }
          while (++k <= fie.iFamSecMax[10]) ;

if (dbGrndArea != 0) {
    dbIEOut[0] = (int)(dbGrndFirstMom/dbGrndArea);
    bIndet[0] = FALSE;
    }
else {
    dbIEOut[0] = 0;
    bIndet[0] = TRUE;
    }
return TRUE;
}
```

FUZZ-C INFERENCE ENGINE

Fuzz-C (Bytecraft, Inc.) differs from the other knowledge-base–inference-engine systems you've seen here. In Fuzz-C, the knowledge base is embedded in the inference engine—it's an integral part of it. The disk contains a file *trktrl.fuz* that was built by the commercial version of Fuzzy Knowledge

Builder™ and has three input dimensions. The file is formatted for embedding in a Fuzz-C inference engine using the Fuzz-C inference engine builder.

The file *trktrl.c,* also on the disk, is the inference engine with trktrl.fuz embedded in it.

MOTOROLA 68HC05 ASSEMBLY SIMPLE INFERENCE ENGINE

This engine is written in assembly language for the Motorola 68HC05 processor, and is in file *ie05.asm* on the disk. This engine performs functions similar to those in the C language engine described earlier in the appendix, but without that engine's context switching ability.

You can test it with test file *trktrlt.f05,* which was built by the commercial version of Fuzzy Knowledge Builder™ from the truck-parking knowledge base mentioned previously. The defined constants are at the end of trktrlt.f05. You can also test it with other .f05 files you may have.

APPENDIX E

OTHER FUZZY ARCHITECTURE

In addition to the three fuzzy architectures detailed and implemented in this book, others exist. This appendix discusses two of them—a fuzzy generalization of the artificial intelligence language OPS5, called FLOPS, and fuzzy state machines.

FLOPS

FLOPS (Fuzzy logic production system) is an inference engine for a rule-based expert system written in the mid-1980s by William Siler and Douglas Tucker (Kemp-Carraway Heart Institute). It uses fuzziness in three ways:

- *Data types.* FLOPS uses a combination of crisp and fuzzy data types. Crisp data types include integers and floating point

numbers. The fuzzy types include numbers, sets, and certainty factors, which represent the degree of truth involved.
- *Fuzzy logic.*
- *Ability to learn.* Because the rules aren't absolute, the program is able to "learn" and "change its mind" when it retraces its steps through earlier rule searches.

FLOPS is a generalized extension of OPS5, a well-known crisp inference engine developed at Carnegie-Mellon University in the 1970s. It also has similarities to the computer language Prolog. The sequential version (which finds and fires one rule at a time) uses forward chaining and emulates backward chaining. A parallel version (which allows firing several rules at once) can also use backward chaining. FLOPS can be used in crisp mode, so that it works like OPS5.

E-MAIL FROM D R. FUZZY
OPS5 (Official Production System version 5) is the best-known of an OPS series. It features If–Then rules that are searched by a method known as *forward chaining*, in which appropriate rules are linked or chained, so that they solve a problem or otherwise lead to a conclusion.

E-MAIL FROM D R. FUZZY
Prolog (programming in logic) is a programming language, developed in the 1970s, that uses logic rules to prove relationships among objects by *backward chaining*. In this method of rule chaining, the program develops a hypothesis, then tests it by working backward through the rules, If the hypothesis holds up when the rule search is concluded, it is considered to be true.

How FLOPS Works

FLOPS is a *production system*, meaning that its rules can be written in any order because all rules are searched before any are selected to be fired. The data in the expert system's knowledge base determines which rules are candidates for being fired. Then FLOPS uses a conflict-resolution algorithm that determines which rules it has the most confidence in. These rules are fired.

When no more rules are available to be fired, FLOPS backtracks through them, remembering what it learned, making the process more efficient. FLOPS also has a set of *metarules*, which are used to create rules from an expert knowledge base.

BADGER—AN ANIMAL GUESSING GAME

As an example, FLOPS includes a guess-the-animal game in which it asks the user to think of an animal and provide clues to its identity when asked by the program. Using this information and its knowledge base of animal characteristics, the program creates a hypothesis, then asks the user of the hypothesis is correct.

For instance, BADGER categorizes animals in fuzzy sets such as Teeny, Small, and Medium. When it asks the user to specify the mystery animal's weight in kilograms, the program then determines confidence levels by making a fuzzy comparison with fuzzy numbers. If this information falls in the set Medium, the program might ask the user whether the animal is a Dog. If the user responds Yes, the game is over. If the response is No, BADGER tries another animal.

As BADGER backtracks in searching for the correct answer, its remembers the user's previous responses and the pathways it has already searched. Fuzziness in the rules involves, for instance, comparing the weight in crisp kilograms to the fuzzy set (.01, 0, .5) that encompasses the size Teeny. When backtracking, the program's crisp logic stores the size with the most likely value. So if the user enters a crisp weight that falls within that fuzzy set, the program will guess

Your animal is possibly Teeny

The size Small governs the fuzzy set (1, 0, .5), and so on.

Other responses from the user are inherently crisp, such as number of legs, type of food, and charisteristics of the skin.

Parallel FLOPS

For parallel operation, FLOPS stores rules in blocks and all rules in a block fire at once. For instance, a response from the user that falls within the categories of Small and Medium causes rules covering both the Small fuzzy set and the Medium fuzzy set to fire at the same time.

If two rules with the identical *then* action try to fire, the program uses logic to determine which rules has the higher confidence level. It then causes that rule to fire.

STATE MACHINES

State machines, which you met in Chapter 6, can also be used as control structures. For example, a state machine can be coupled with a rule base to determine which rules will be fired and in which sequence. If a state machine is capable of being in one of three states, each state can govern a different group of rules or even a different rule base.

Crisp State Machine

A crisp state machine is in one state at a time. If three states—say, A, B, and C—are available, the crisp state machine is in state A or state B or state C. The transition from one state to another is triggered by an incoming crisp event, so that the new state depends on the interaction of the old state and the event, such as,

> Event + State A results in State B
> Event + State B results in State C
> Event + State C results in State A

Differing events might also result in different new states, for instance,

Event A + State A results in State B
Event B + State A results in State C
Event A or Event B + State C results in State A

and so on.

In a control system, each state may be linked to an action, so that

Event A + State A results in State B, which causes Action B
Event B + State A results in State C, which causes Action C
Event A or Event B + State C results in State A, which causes Action A

Fuzzy State Machine

If you generalize to a fuzzy state machine, the machine can have degrees of membership in the current state. This is similar to membership in a fuzzy set. As with fuzzy set membership, just because the state machine is partially in a state, it isn't necessarily partially in another state. In addition, the events are also fuzzy.

In orthodox fuzzy operations, such a system would ultimately degrade, because the degree of membership in a new state can never be greater than the degree of membership in the previous state. And it isn't possible for an event to simply strengthen or weaken the existing state, rather than change the state. There are several ways to get around this problem:

- Ignore the strength of the transition event. Make the degree of membership in the new state depend only on the degree of membership in the existing state.
- Ignore the value of existing state. Make the degree of membership in the new state depend only on the causal event's degree of membership. This allows the event to strengthen or weaken the existing state.
- Require the degree of membership in the new state to equal 1.
- Make the degree of membership in the new state dependent on a disjunction (v): the degree of membership in the existing state *or* the degree of membership in the event. This allows

only the strengthening of the current state (but not its weakening).

- Make the degree of membership in the new state dependent on the conjunction (∧): the degree of membership in the current state AND the degree of membership in the triggering event. This is the solution recommended by fuzzy logic expert David L. Brubaker (Huntington Advanced Technology).

Putting a Fuzzy State Machine into Operation

Implementing a fuzzy state machine is a five-step procedure.

Identify the States

In a control system, there are typically three types of states: those in which the system operates during startup, the states that are typical of normal operation, and the states that exist during abnormal operation. Unlike an expert system, where fuzzy sets overlap, a fuzzy state machine can be in only one state at a time.

Instead of input speeds, such as Fast or Slow, the fuzzy state machine might have states named Stopped, Accelerating, Decelerating, and Constant Velocity.

Create Membership Functions for Each State

The degree of membership can be a function of the action represented by the state. For example, in a state named Accelerating, the degree of membership might be a function of the rate of acceleration.

Determine the Actions for Each State

If the control system is rule-based, each action is the result of rules firing. Since this is a fuzzy system, the degree to which any rules are fired can be related to the degree of membership in the existing state.

Identify the Triggering Events

All inputs to the system should be examined for their ability to cause state transitions.

Determine the Membership Function for the Next State

Events, too, may be fuzzy. If the recommended conjunction of fuzzy event and fuzzy state is used, the next state's degree of membership must be determined as a function of the degree of membership of the event and the degree of membership in the existing state.

The Rules and the Inference Method

If the system is small and the rule base is relatively simple, forward chaining can be used to search the entire base each time to select the rules to be fired. In a more complex system, each state may have a series of rules linked to it, and only these are searched during the inference process.

Going in the opposite direction, the rule firing itself may actually be the event that triggers the state transition.

BIBLIOGRAPHY

ARTICLES

Brubaker, David I. (Huntington Advanced Technology). Fuzzy-Logic Basics: Intuitive Rules Replace Complex Math. *EDN*, June 18, 1992, p. 111.

Brubaker, David I. (Huntington Advanced Technology). *Huntington Technical Brief*, monthly newsletter, Aug. 1990–Aug. 1993.

Kosko, Bart. Fuzzy Logic. *Scientific American*, July 1993, p. 76.

New York Times. "To Be Precise," editorial. Aug. 26, 1993, p. A14.

O'Hagen, Michael. A Fuzzy Decision Maker. *Proc. Fuzzy Logic '93* (*Computer Design* Magazine), p. M313.

O'Hagen, Michael. Fuzzy Decision Aids. *Proc. 21st Annual Asilomar Conf. on Signals, Systems, and Computers* (IEEE and Maple Press, Pacific Grove, CA), 2, p. 624.

276

O'Hagen, Michael. Aggregating Template or Rule Antecedents in Real-Time Expert Systems with Fuzzy Logic. *Proc. 22nd Annual Asilomar Conf. on Signals, Systems, and Computers* (IEEE and Maple Press, Pacific Grove, CA), 2, p. 681.

O'Hagan, N.K., and O'Hagan, M. Decision-making with a Fuzzy Logic Inference Engine. *Proc. Applications of Fuzzy Logic Technology*, Sept. 1993, Society of Photo-Optical Instrumentation Engineers, p. 320.

Pinder, Jeanne B. "Fuzzy Thinking Has Merits When It Comes to Elevators," *New York Times*, Sept. 22, 1993, p. C1.

Schwartz, Daniel and George J. Klir. Fuzzy Logic Flowers In Japan. *IEEE Spectrum*, July 1992, p. 32.

Self, Kevin. Designing WIth Fuzzy Logic. *IEEE Spectrum*, November 1990, p. 42.

Stubbs, Derek. *Sixth Generation Systems,* monthly newsletter. P.O. Box 155, Vicksburg, MI 49097.

Williams, Tom.Fuzzy Logic Is Anything But Fuzzy. *Computer Design*, April 1992, p. 113.

BOOKS

Axelrod, Robert. *Structure of Decision: the Cognitive Maps of Political Elites.* Princeton, NJ: Princeton University Press, 1976.

Bezdek, James C., and Sankar K. Pal, eds. *Fuzzy Models For Pattern Recognition.* New York: IEEE Press, 1992.

Brubaker, David I. (Huntington Advanced Technology). *Introduction to Fuzzy Logic Systems.* The Huntington Group, 883 Santa Cruz Ave., Suite 27, Menlo Park, CA 94025-4669. 415/325-7554.

Cognizer Co. (Lake Oswego, OR). *Neural Network Alamanac 1990–1991.*

Craig, J.J. *Introduction to Robotics: Mechanics and Control*, 2nd ed. Reading, MA: Addison-Wesley, 1989.

Driankov, Dimiter, Hans Hellendoorn, and Michael Reinfrank. *An Introduction to Fuzzy Control*. New York: Springer-Verlag, 1993.

Jamshidi, Mohammad, Nader Vadiee, and Timothy J. Ross, eds.*Fuzzy Logic and Control. Software and Hardware Applications*. Englewood Cliffs, NJ: PTR Prentice-Hall, 1993.

Klir, George J., and Tina A. Folger. *Fuzzy Sets, Uncertainty, and Information*. Englewood Cliffs, NJ: Prentice-Hall, 1988.

Kosko, Bart. *Neural Networks and Fuzzy Systems*. A Dynamical Systems Approach to Machine Intelligence. Englewood Cliffs, NJ: Prentice-Hall, 1992.

Kosko, Bart. *Fuzzy Thinking. The New Science of Fuzzy Logic*. New York: Hyperion, 1993.

Leigh, William E., and Michael E. Doherty. *Decision Support and Expert Systems*. Cincinnati: South-Western Publishing Co., 1986.

McNeill, Daniel, and Paul Freiberger. *Fuzzy Logic. The Discovery of A Revolutionary Computer Technology—and How It Is Changing Our World*. New York: Simon & Schuster, 1993.

Siler, William and Douglas Tucker. *FLOPS User's Manual*. Birmingham, AL: Kemp-Carraway Heart Institute (1600 North 26th St.), 1986.

Thro, Ellen. *The Artificial Intelligence Dictionary*. San Marcos, CA: Microtrend Books, 1991.

Yager, R.R., S. Ovchinnikov, R.M. Tong, and H.T. Nguyen, eds. *Fuzzy Sets and Applications: Selected Papers by L.A. Zadeh*. New York: John Wiley & Sons, 1987.

Zadeh, Lotfi, and Janusz Kacprzyk, eds. *Fuzzy Logic for the Management of Uncertainty*. New York: John Wiley & Sons, Inc. 1992.

Zimmermann, Hans-J. *Fuzzy Sets, Decision Making, and Expert Systems*. Boston: Kluwer Academic Publishers, 1987.

CONFERENCE PROCEEDINGS

Computer Design Magazine. *Proc. Fuzzy Logic '93.*

Fuzzy Logic Systems Institute, Inc. *Proc. 2nd International Conference on Fuzzy Logic and Neural Networks [IIZUKA '92],* 2 vols. New York: Fuzzy Logic Systems Institute, 1992.

IEEE Neural Networks Council.*Proc. Second IEEE International Conference on Fuzzy Systems, Mar. 28–Apr. 1, 1993. (FUZZ-IEEE 93),* 2 vols.

Kyusu Institute of Technology. *Proc. International Conf. on Fuzzy Logic and Neural Networks [IIZUKA '90].* Especially:

E. Akaiwa *et al.* Hardware and Software of Fuzzy Logic Controlled Cardiac Pacemaker, p. 549.

J.L. Castro *et al.* On the Semantic of Implication, p. 719.

M. Delgado *et al.* The Generalized "Modus Ponens" with Linguistics Labels, p. 725.

T. Hakata and J. Masuda. Fuzzy Control of Cooling System Utilizing Heat Storage, p. 77.

S. Kageyama *et al.* Blood Glucose Control By a Fuzzy Control System, p. 557.

R. Lopez de Mantaras *et al.* Connective Operator Elicitation For Linguistic Term Sets, p. 729.

E.H. Ruspini. Similarity-Based Interpretations of Fuzzy Logic Concepts, p. 735.

L. Valverde and L. Godo. On the Functional Approach to Approximate Reasoning Models, p. 739.

T. Watanabe *et al.* "AI and Fuzzy"-Based Tunnel Ventilation Control System, p. 72.

INDEX

— M —

— N —

— O —

— P —

— Q —

FUZZY THOUGHT AMPLIFIER™

The FUZZY THOUGHT AMPLIFIER helps you model the real world. The FUZZY THOUGHT AMPLIFIER provides an intuitively helpful interface for modeling, capturing and exercising thought models..

- ☑ The FUZZY THOUGHT AMPLIFIER allows up to twenty five conceptual states and six hundred twenty five interconnecting causal events.
- ☑ The FUZZY THOUGHT AMPLIFIER provides a graphical interface for creation, placement and viewing of the conceptual states and the interconnecting causal events in the classical "directed graph with feedback" visually intuitive format..
- ☑ The FUZZY THOUGHT AMPLIFIER provides for spreadsheet editing and viewing of the concept state names and activations and of the causal event names and weights. Each spreadsheet cell is dynamically updated after a map step.
- ☑ The FUZZY THOUGHT AMPLIFIER provides single step, multiple step and continuous run control. At each step the states are defined anew as a result of the causal vector addition connection to the previous states through the event weights. The resulting dynamic graph may terminate in a static, limit-cycle or chaotic condition. Just like reality.
- ☑ The FUZZY THOUGHT AMPLIFIER allows complete visual control of the inferencing squashing function. This function monotonically squashes the causal addition vector into the state activation limits. Limit cycle effects are set through an inference roughness control.
- ☑ The FUZZY THOUGHT AMPLIFIER provides three different run time representation of the state activations: the directed graph, the spreadsheet listing and as a color bar chart. These may all be viewed simultaneously and dynamically.
- ☑ The FUZZY THOUGHT AMPLIFIER includes the following intelligent tools:
  - ♦ The FUZZY THOUGHT MELDER provides combine and arrange functions that allows combining two maps of differing credibilities to produce a third melded and arranged map.
  - ♦ The FUZZY THOUGHT TRAINER provides a training function that automatically adjusts the existing event weights to produce the existing defined set of state activations.
  - ♦ The FUZZY THOUGHT OBSERVER provides an observer function that provides automatic stopping of the map evolution on limit cycles up to 10 steps deep.
- ☑ The FUZZY THOUGHT AMPLIFIER is a Windows program. All windows support functions are available. The various tools and states may easily be clipped and printed.

Product includes software on 3.5 inch high density disk, users manual and registration form. Software includes the FUZZY THOUGHT AMPLIFIER and the examples as outlined in the manual. The FUZZY THOUGHT AMPLIFIER and examples requires MS Windows 3.1, 386 or better PC with mouse and VGA or better.Cost is $195, $15 S/H and appropriate California state sales tax.

FUZZY SYSTEMS ENGINEERING
12223 WILSEY WAY, POWAY, CA 92064
Phone/Fax (619) 748-7384

FUZZY KNOWLEDGE BUILDER™

The Fuzzy Knowledge Builder helps you design fuzzy systems. The Fuzzy Knowledge Builder provides an intuitively helpful interface for capturing the expert judgments needed to build any fuzzy control or transform system.

☑ The *Fuzzy Knowledge Builder* allows up to eleven input dimensions and two output dimensions in the fuzzy control surface design. Each dimension may be described by up to eleven fuzzy sets.

☑ The *Fuzzy Knowledge Builder* provides a two dimension activation matrix type display of the fuzzy system rules for editing of those rules. Rules editing is by simple mouse point and clicking. Full text description of the rule in focus is displayed.

☑ The *Fuzzy Knowledge Builder* provides piece-wise editing of the fuzzy sets through simple mouse point, click and drag operations.

☑ The *Fuzzy Knowledge Builder* outputs the expert fuzzy systems knowledge base in an include or data file for use in your application. Many languages are supported such as C and assembly.

☑ The *Fuzzy Knowledge Builder* includes many supportive functions including:

  ☑ *Knowledge Action Tester™* allows static testing of the fuzzy estimation surface at any time in the design cycle.

  ☑ *3D Viewer™* provides 3D viewing of the fuzzy parametric estimation surface.

  ☑ *Gradient Viewer™* provides contour mapping of the fuzzy surface.

  ☑ *Profile Viewer™* provides orthogonal profiles of the surface.

  ☑ *Knowledge Copy™* will copy rule groups to multiple rule groups, from rule matrix slice to rule matrix slice and from fuzzy set group to fuzzy set group..

  ☑ *Knowledge Mix™* will fill the rule cells with random values.

  ☑ *Knowledge Grade™* constructs hyper-dimensional gradients on the rules.

  ☑ *Knowledge Automata™* smoothly extrapolate rules between a few defined and fixed hyper dimensional rules.

Product includes software on 3.5 inch high density disk, users manual and registration form. Software includes the Fuzzy Knowledge Builder, the examples as outlined in the manual and an example executable fuzzy system, with source, that uses a Fuzzy Knowledge Builder output include file. The Fuzzy Knowledge Builder and examples requires MS Windows 3.1, 386 or better PC with mouse and VGA or better. **Cost is $295, $15 S/H and appropriate California state sales tax. See order form.**

FUZZY SYSTEMS ENGINEERING
223 WILSEY WAY, POWAY, CA 92064
Phone/Fax (619) 748-7384

FUZZY DECISION MAKER™

The Fuzzy Decision Maker helps you make complex decisions. The Fuzzy Decision Maker provides an intuitively helpful interface for breaking apart and capturing the simple subjective judgments needed to make complex large decisions involving here and now conditions (constraints); the future desired conditions (goals); and the various optional paths for getting from constraints to goals (alternatives).

☑ The Fuzzy Decision Maker allows up to twenty goals, twenty constraints and forty alternatives in the decision process.

☑ The Fuzzy Decision Maker provides an intuitive graphical interface for simple ranking of the goals importance and constraints importance in the decision.

☑ The Fuzzy Decision Maker provides an intuitive graphical interface for simple ranking of the satisfactions found in the alternatives for each goal and constraint.

☑ The Fuzzy Decision Maker also provides an alternate numerical interface for percentage ratings of these importances and satisfactions.

☑ The Fuzzy Decision Maker provides an entry of your optimism or pessimism on the decision process. This can greatly yet appropriately influence the decision results.

☑ The Fuzzy Decision Maker provides a color bar chart interface for absolute ranking of the alternatives at the conclusion of the decision. The highest bar is the best choice.

☑ The Fuzzy Decision Maker conclusion bar chart may show the contributing parts of the decision as color stacks in the bar chart.

☑ The Fuzzy Decision Maker can print out a formatted report on your decision scenario and the decision results.

☑ The Fuzzy Decision Maker is a Windows program. All windows support functions are available.

Product includes software on 3.5 inch high density disk, users manual and registration form. Software includes the Fuzzy Decision Maker and the examples as outlined in the manual. The Fuzzy Decision Maker and examples requires MS Windows 3.1, 386 or better PC with mouse and VGA or better. Cost is $195, $15 S/H and appropriate California state sales tax. See order form.

Fuzzy Logic, Inc.
1160 Via España
La Jolla, CA 93037
(619)456-2634 Ph. & FAX
E-mail:fuzzyli@crash.cts.com

FUZZY SYSTEMS ENGINEERING
12223 WILSEY WAY, POWAY, CA 92064
Phone/Fax (619) 748-7384

FUZZY TOOLS ORDER FORM

YES! I would like to order __ Fuzzy Knowledge Builders x $295 each = _____

and __ Fuzzy Decision Makers x $195 each = _____

and __ Fuzzy Thought Amplifiers x $195 each = _____

Calif. customers add 7.5% sales tax: _____

Shipping: $15 U.S. or Canada, $50 overseas: _____

TOTAL = _____

Shipping Address: (Please provide a street address. No P.O. boxes. Please print.)

Name:_____ Company:_____

Address:_____

City:_____ State/Country:_____ Zip:_____

Phone:_____ Fax:_____

Method of Payment: (Check one)
____ Check/money order (US$, US bank)

____ P.O.#(Approval required):_____

____ MasterCard   ____ Visa   ____ American Express

Credit Card #:_____ Exp. Date:_____

Cardholder's name:_____

Return to: Fuzzy Systems Engineering   FREE FSE T-SHIRT
            12223 Wilsey Way   WITH EVERY ORDER!
            Poway, CA 92064   Circle Shirt Size: S M L XL XXL
            Phone 619-748-7384
            Fax 619-748-7384

FILE INSTALLATION PROCEDURE

The files are in compressed form on the floppy disk and will be expanded automatically during the installation procedure.

1. With Windows active, insert the floppy disk in the drive and select that drive. Click on the **instalit.exe** file.

2. A dialog box will ask if you want to proceed with the installation. Click on OK.

3. A dialog box will ask whether the default directory is OK. Click on OK.

4. Installation will proceed, placing the files in the **fuztools** program group.

5. Open any file by double-clicking on its icon.